Cosmic Company

The Search for Life in the Universe

Is there life elsewhere in the Universe? What might it be like and how will we ever find it? *Cosmic Company* ponders the possibility of aliens visiting the Earth, as well as what it would mean if we were to pick up a signal from the Cosmos that would prove we're neither alone, nor the smartest creatures in creation. It explains why scientists think life might be plentiful on other worlds, and how we might get in touch. Containing a thorough overview of the science and technology behind the search for life, the book highlights current and future space missions and research, which are aiming to answer some of the greatest questions mankind has ever asked. This easy-to-read book, by two experienced writers of popular astronomy, is suitable for anyone who ever wondered whether there's anybody out there...

SETH SHOSTAK is a senior astronomer at the SETI (Search for Extraterrestrial Intelligence) Institute in California, a research organization that runs the world's most sensitive search for extraterrestrial intelligence. Seth has a degree from Princeton University, and a doctorate in astronomy from the California Institute of Technology. For years he conducted radio astronomy research on galaxies, and worked at both national observatories and universities in the USA and Europe. Seth has written several hundred popular articles on various topics in astronomy, technology, film, and television, as well as more than fifty research papers. He lectures frequently at the California Academy of Sciences and elsewhere, and for the last four years has been a Distinguished Speaker for the American Institute of Aeronautics and Astronautics.

ALEX BARNETT is a well-known figure in the science center, planetarium and media world, particularly for public and educational programs involving space and astronomy. After completing her astrophysics degree, Alex worked at the BBC in many productions for both radio and television. She presented the BBC's Space and Astronomy program *Final Frontier* for two years and is a frequent guest on science programs. Alex spent six years as the vision behind the National Space Centre, a new visitor attraction that opened in Leicester, UK in June 2001, and led the team that created an award-winning exhibition and Space Theatre. She is currently CEO and Executive Director of the Chabot Space and Science Center in Oakland, California.

Cosmic Company

The Search for Life in the Universe

SETH SHOSTAK *SETI Institute*
ALEX BARNETT *Chabot Space and Science Center*

PUBLISHED BY THE PRESS SYNDICATE OF THE UNIVERSITY OF CAMBRIDGE
The Pitt Building, Trumpington Street, Cambridge, United Kingdom

CAMBRIDGE UNIVERSITY PRESS
The Edinburgh Building, Cambridge CB2 2RU, UK
40 West 20th Street, New York, NY 10011–4211, USA
477 Williamstown Road, Port Melbourne, VIC 3207, Australia
Ruiz de Alarcón 13, 28014 Madrid, Spain
Dock House, The Waterfront, Cape Town 8001, South Africa

http://www.cambridge.org

First published 2003

Printed in the United Kingdom at the University Press, Cambridge

Typefaces Swift 11/14.5 pt. and Quadraat headliner *System* LaTeX 2ε [TB]

A catalog record for this book is available from the British Library

ISBN 0 521 82233 5 hardback

Contents

Acknowledgements

We would like to express our appreciation to the creative team at the National Space Centre – Annette Sotheran, Andy Gregory, Roger Jones, Max Crow, and Helen Osbourn – for their help in creating and supplying many of the images in this book.

Introduction. Could there be aliens?

This is a book about the quest for other beings.

Do aliens exist? Could it be that, as you read these words, hordes of thinking creatures are living out their lives on worlds we have never seen? Are untold trillions of civilizations sprinkled throughout the enormous spaces of the Universe?

This is an old question, and one that has probably been asked ever since humans were clever enough to wonder about such things. In the past, the only way to get an answer was to make one up. And people did. The ancient Greeks believed that the stars and planets of the night sky were homes to gods and other men. For them, the entire Cosmos was populated. The Europeans of medieval times weren't quite so inclined to believe that the heavens were filled with thinking aliens, for that, they feared, would make Earth and its peoples less important to God. In the last few centuries, this argument has been occasionally turned around: why would God construct such an enormous Universe if only one small planet was to be inhabited? Isn't that like building a 40-room mansion, and camping out in the hallway? A waste of space?

Today, polls show that roughly half the population of Europe and America believes that extraterrestrial beings are out there.[1] It's probable that many of them have come to this conclusion after watching countless film and television dramas in which creatures from other worlds either come to Earth to take over our planet, or engage in space battles with our descendants. (Very few of the cinema aliens are friendly, but that's because they're much more useful to the story as pitiless bad guys.) However, just because extraterrestrials are plentiful on the silver screen doesn't mean that they're plentiful in space. Talking ducks and chatty rabbits are common in the movies, but not in the countryside.

[1] A 2001 Gallup poll taken in the United States revealed that 71 percent of the population believes that the American government is covering up evidence of alien visitation.

To borrow a well-known phrase from the late Douglas Adams, "The Universe is Big, really Big."

A popular view of alien visitation, but these little gray guys are made of silicone rubber.

So despite the fact that we're more than two thousand years beyond the ancient Greeks, we can't do better than make up answers to the question of whether we're alone. If we knew the truth, it would be enormously important to us. For example, if we were to find that our Galaxy is swarming with advanced societies, that would at least give us some perspective on just how significant we are in the scheme of things. There's also the possibility that we could learn wonderful things from advanced civilizations in other star systems, and possibly leap into a greatly improved future. On the other hand, if it turns out that Earth is the only world to be inhabited by thinking beings, then we would have to face up to the daunting fact that the future of the Cosmos depends entirely upon us.

So you see that the question of alien existence is more than simply an interesting conversation topic to raise at your next social gathering. And, of course, it always was. But the difference today, as the twenty-first century shifts into second gear, is that we have some chance of answering that question by observation, and not just with someone's opinion. Unlike the ancient Greeks, the medieval Europeans, or even the accomplished scientists of Victorian times, we have the knowledge and technology to actually prove the existence of aliens. That isn't to say that we *will* succeed, or if we do it will happen soon, but it might.

Things to come

Our search for intelligent life beyond Earth will take us to the stars. But which ones are worth the bother? Where, in the endless desert of space, might we hope to find suitable planets – the oases where thinking beings

Are there worlds out there in the depths of space where life might be lurking?

could bloom? We know that life on Earth is tough: it can survive in conditions harsher than drain cleaner. But what are the limits for life in general?

Even assuming the aliens are out there, what would they be like? Should we expect little gray guys with almond eyes and no hair? Could they have a thousand legs and be no bigger than a rat? Or will the extraterrestrials be completely synthetic: advanced computers that live forever and dwarf earthly intelligence?

In addition to trying to answer these questions, we'll also discuss the actual hunt for thinking aliens: the experiments known as SETI, the Search for Extraterrestrial Intelligence. Will such searches ever be successful, and if so when? What about UFOs and other phenomena that some people offer as proof of aliens on Earth? And finally, suppose we do find Jo Alien? What will that mean to us?

This book is about a truly exciting adventure: the serious attempt to learn if we are alone in the Cosmos. For thousands of years, humans have explored the world in search of new places and other cultures. Now, finally, we are exploring the realm of the stars.

Home sweet home and the only one we currently have. But is Earth a highly unusual planet?

CHAPTER 1

Habitats for life

One of the most compelling arguments for life elsewhere is that Earth is not special.

Well, of course *you're* special, and so are the people you know and the places you go. But that's not the point. In discussing whether aliens could exist in the depths of space, we first need to ask whether there's something special about our planet. Is Earth constructed of unusual materials or located in some particularly friendly part of the Universe? As it turns out, most astronomers think the answer to this question is "no."

You might be surprised at their confidence, given that the Cosmos is big enough to boggle your brain. However, before we explain why we can be so sure that even the most distant objects we can see in our telescopes are made of the same sorts of materials as your local neighborhood, we should first briefly describe the nature of galaxies, stars, and planets: the familiar objects of the Cosmos.

Planets, stars, and galaxies

Just about everyone has a good idea of the difference between stars and planets. Stars, like the Sun, are huge balls of hot gas massive enough to crush their innards to enormous temperatures and pressures. This causes the central cores of stars to undergo non-stop nuclear reactions, similar to those that take place within a hydrogen bomb. These reactions generate the energy that causes the stars to shine. Smaller, much cooler bodies that orbit stars are known as planets, and planets *don't* shine. Every schoolchild knows the nine planets of our own Solar System, of which five are near enough to Earth to be visible to the naked eye (Mercury, Venus, Mars, Jupiter, and Saturn). These cool worlds (and their moons) were made from the "leftover" material that accompanied the birth of our Sun. Until very recently, the only planets we knew about were those of our Solar System, but

The start of it all

The Universe began with a massive bang, about 14 billion years ago. Most people know this, and many of them imagine the Big Bang as an enormous explosion – a sudden flash of light – that took place in a huge and dark emptiness that had been sitting around forever, waiting for something to happen.

In fact, the Big Bang didn't erupt into a large, empty room. There was no such thing as "room" before the explosion occurred. Space was created in the explosion, and when the Big Bang happened, the blast occurred everywhere at once. "Everywhere" wasn't very large at first. One hundred-million-million-million-million-millionths of a second after the bang, the entire Universe was roughly the size of a grapefruit. But it grew quickly, and within a mere three minutes – the time required to boil an egg – the basic building blocks of the Cosmos had been put together.

If you could have visited the Universe at this time, you would have found an intensely hot gas of tiny particles, zipping around in a blinding sea of light. The light was endlessly bouncing off these particles, and unable to travel very far in a straight line. For a few hundred thousand years, this picture didn't change much. But as the newborn Universe cooled, the particles were able to group together to form atoms of hydrogen and helium – the two lightest elements. With the smallest particles removed from the fray, the light was able to travel long distances and escape the local scene – and rather quickly the Universe turned dark.

For about a billion years, there was only blackness, and the silent swirl of clouds of hydrogen and helium. Eventually and inevitably, some of the clouds of gas collapsed under their own gravity, making clumps of stars. But these stars were not randomly spread throughout the growing Universe, like dust in a room. Instead, they were assembled in giant neighborhoods we now call galaxies.

The gloom of a billion years was gone, broken by the pinpoint lights of countless stars. The familiar face of the Universe had emerged.

This image shows the Universe when it was rather younger. Some of these galaxies are mere pups.

astronomers are now finding that planets are common around other stars too. Not only that, but many young stars seem to be surrounded by the dusty disks that we believe are the material from which planets are born. This bodes well for the possibility of extraterrestrial life, for we know that biology won't arise on stars (entirely too toasty!), but there's a much better chance for life to get a foothold on the cool, wet landscapes of planets. Obviously, the more planets there are, the greater the opportunities for life.

You can see a few hundred stars with your naked eye on a clear, dark night (unless you live in a city or suburb where the glare from street lamps ruins the view). If you use a pair of binoculars or a small telescope, you can make out thousands of stars. In fact, if you could view all these stars from a great distance, you would notice that they are arranged in a large, flat, spiral-shaped structure called a galaxy. The star count in our galaxy – the Milky Way – is several hundred billion, or roughly the number of grains of sand it would take to fill a dump truck. There are about a hundred billion *other* galaxies visible to our best telescopes, each with a similar number of stars. So the total tally of stars we can see is approximately equal to the number of sand grains in the Sahara desert, a fact you can use to impress other people while waiting for the bus.

That's a lot of stars. And as we've remarked, they're all built of the same stuff. How do we know this? After all, our spacecraft have collected precious little material from the Universe: 382 kg of Moon rock, to be specific. We also have some meteorites from the outer Solar System and a few small hunks of rock that were blasted off Mars in the distant past. All these are of great interest, but are merely debris from nearby objects – samples from our cosmic backyard. However, despite the fact that we don't have specimens from far-off stars and galaxies, we can still be confident of their composition. That's because what we *do* have – the light they emit – tells us what they're made of. The gases in the outer layers of stars influence the light they radiate into space. Every type of gas will absorb certain colors, or wavelengths, of light. If we break up the starlight entering our telescopes into a rainbow, we can see that some wavelengths – certain very specific colors – are missing. The rainbow resembles a colored barcode. We can link these dark absorption lines to specific chemical elements such as hydrogen, helium, calcium, and iron. It's like using fingerprints to identify individual people.

Close up, our Sun shows what a violent place it really is. Needless to say, life could not arise on its surface.

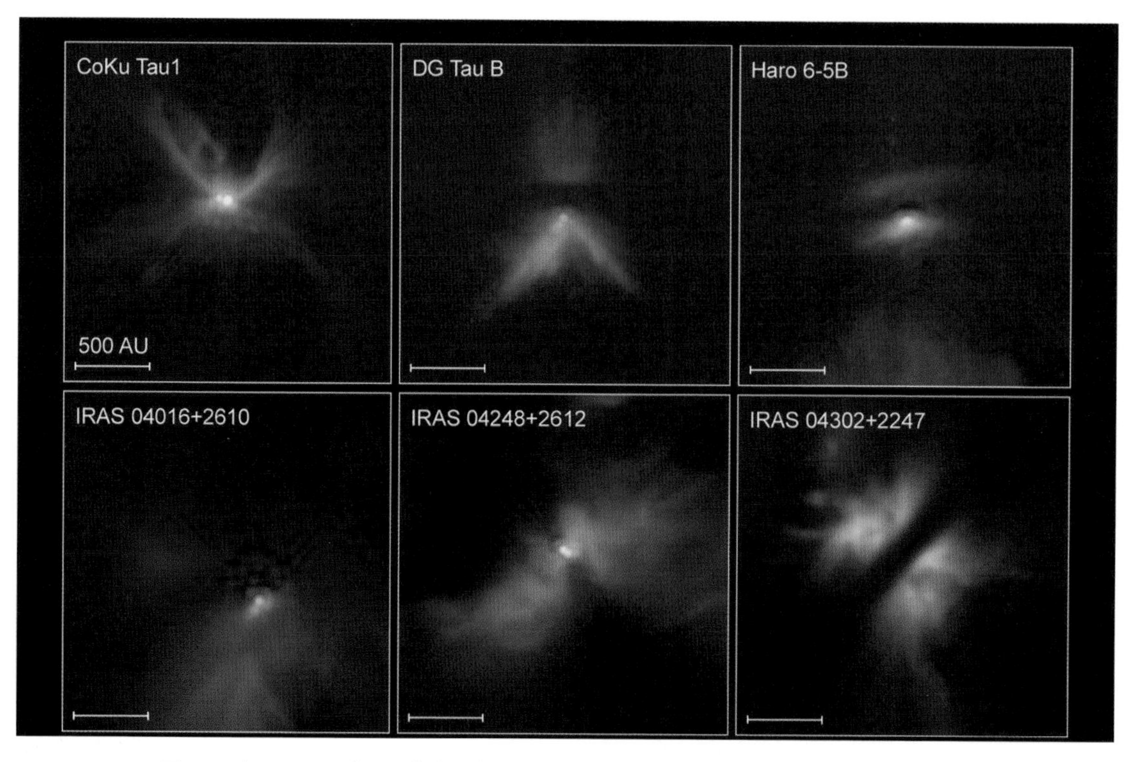

These close-ups of stars being born show that dusty disks, the kind from which planets form, should be quite common. The bars at the bottom indicate a distance that is 500 times that between the Earth and the Sun.

When we apply this type of analysis to the light from stars in even the farthest galaxies, we find the same, familiar elements that make up you, your house, and Mars.

Good suns

It is an amazing circumstance, and one of the great achievements of astronomy, to know that stars everywhere are all made of the same materials (although, admittedly, they might have slightly differing amounts of each element).[1] In addition, the very fact that we can successfully analyze

[1] In the middle of the nineteenth century, the French philosopher Auguste Comte wrote that we would never know anything about the composition of the stars, given the fact that they were so far away. As it turns out, only a few years after Comte's death, the German physicists Gustav Kirchoff and Robert Bunsen developed the spectroscope for analyzing light, and proved the famous philosopher wrong.

If we could see our Milky Way galaxy from the outside it might look like this. This is NGC4414.

starlight means that the laws of physics and chemistry are the same everywhere. You should find this a great relief, as it means you won't have to take lessons in these subjects again if you move to another galaxy. On the other hand, it also suggests that whatever has occurred in this part of the Universe – for example, the development of intelligent life – could happen anywhere.

Well, OK, maybe not *anywhere*. While there are 10,000,000,000,000,000,000,000 stars in the known Universe, not all of them will host great neighborhoods

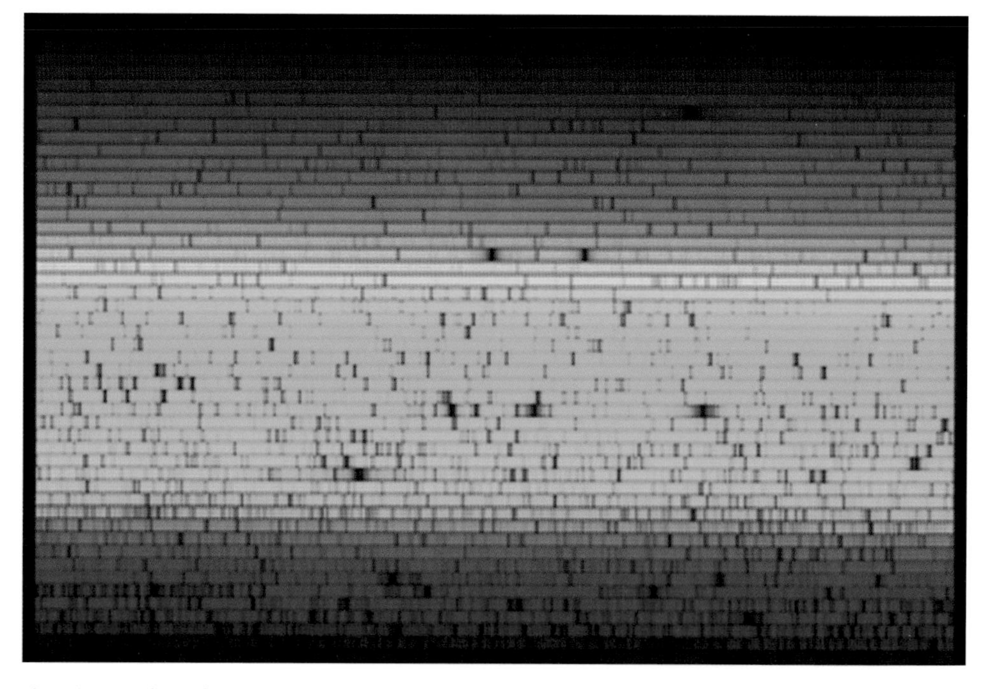

The "bar code" of our Sun. Each dark line is due to a particular chemical element in the Sun.

for life. Stars come in a wide variety of sizes, but for life, bigger is probably not better. Arcturus, Betelgeuse, Canopus, and many of the other bright stars that are familiar to anyone who's glanced at the night sky are all giants. They are several times larger than the Sun (which is about 1.5 million kilometers across), and *thousands* of times brighter.

It's nice to be bright, but such big stars burn through their nuclear fuel much faster than small ones. Of course, being larger, they have more fuel to burn, but this hardly compensates for their fierce energy use. Big stars typically race through their fuel supplies in a few tens of millions of years. Then they end it all by blowing up in a massive explosion: a supernova. They have a brilliant life, but a brief one.

So why is this an issue for the appearance of life? Well, we believe that biology might take hundreds of millions of years (or more) to get underway, and if this is the case, then a big star's lifetime is simply too short to allow complex life to brew up on any surrounding planets. So we probably won't find aliens hanging out around giant stars.

Fortunately, that doesn't matter much. It's true that nearly all the naked-eye stars are giants. But that's not because such weighty stars are plentiful; it's just that they're easier to see because they're so bright. If you do a census of *all* stars, you'll find that only one in a hundred is a heavyweight. Among the others, roughly one in ten is approximately Sun-size, and the overwhelming majority – nearly nine out of ten – are dwarf stars.

Can these common, smaller stars host habitats for life? To answer this question, we need to take a quick look at what we know about life. We don't understand in detail how biology got started on Earth (let alone on any other planet), but one ingredient we believe was essential is liquid water. Water is more than just a cool drink on a warm day. Life, when you get right down to it, is chemistry, and chemistry consists of reactions between various compounds. If organic compounds are floating in a liquid (such as happens within a living cell), they can more easily meet and react. A cell without liquid would be like a party where everyone was forced to sit in a chair all night, unable to drift around the room. Social contacts would be limited. In addition to encouraging reactions, liquids are also useful for bringing food to cells, and for carrying waste products away. So a fluid is required to keep cells functioning, and is presumably also necessary to get the molecules of life together in the first place.

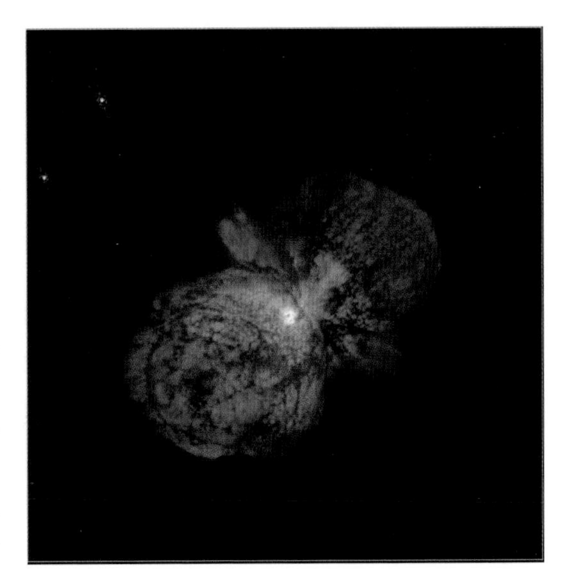

Eta Carina is a very unstable star of the kind that we probably wouldn't expect to support life. Biology would never have time to get started before being wiped out.

But does that fluid have to be water? It's possible that some other liquids could promote the chemistry of life, but let's face it, water has some handy properties and is abundant (can you think of any other common, naturally occurring liquids on Earth?). Thanks to its peculiar molecular structure, water is an excellent solvent, which is very useful in breaking down chemicals so they can react with one another. It also has the (unusual) feature that, when frozen as ice, it becomes less dense. In other words, ice floats. If this wasn't true, then any lake or sea with life would

The position of the habitable zone around a star depends on the type of star.

be sterilized in winter, as ice forming at the surface would drop to the bottom, allowing more surface water to freeze... until the entire lake or ocean was solid and dead.

Astronomers are willing to make large bets that liquid water is the essential ingredient for life. Consequently, on planets where the temperatures are always below water's freezing point, the outlook for biology is poor. If our Solar System is any guide, such chilled-out worlds may be quite common; indeed, the majority of planets might be lifeless deep-freezes, permanently shrouded in deathly white gowns of ice.

At the other extreme, we expect to find many planets hovering so close to their suns that surface temperatures are above boiling. These hellish habitats might be quite dramatic in appearance, but are unlikely places for life to either appear or survive. Not only would these worlds soon boil away their oceans, but the high temperatures would quickly break apart any of life's complex molecules.

Planets that are too hot or too cold aren't going to get much attention from anyone looking for biology beyond Earth. Instead, somewhat like Goldilocks' quest for a good bowl of porridge, these researchers hope for worlds that are "just right" for liquid water. And this is where the smaller stars lose out.

The fact that dwarf stars are smaller than the Sun – a star we know hosts habitats for intelligent life – is not an issue. What *is* an issue is the fact that they're much dimmer. They typically pump out only a tiny fraction of the energy produced by our Sun, and it is this fact that reduces the chances that such stars are homes to life. Consider the following: in our Solar System an Earth-like planet orbiting at any distance greater than Venus and less than Mars passes the "Goldilocks test," and could have liquid water on its surface. Such a planet would get enough sunlight to keep water from freezing, but not so much as to cause it to boil. In other words, there's a "habitable zone" in our Solar System – a doughnut of space around the Sun where planets are neither too frigid nor too toasty to have liquid water, and the thickness of that doughnut is about 100 million kilometers.

But for a dwarf star whose dim glow is only 1 percent that of the Sun, this doughnut would be a lot smaller and thinner, with a width of only 10 million kilometers. So the chance that a random planet would be in the habitable zone of a dwarf star is only about one-tenth what it would be around a Sun-like star. This is the bad news. But there's also good news. Even though dwarf stars are unlikely places to find Jo Alien, there are a lot of these puny suns around. At least some of them will beat the odds and have habitable worlds, and we can't really rule out the dwarfs as worthy stellar homes.

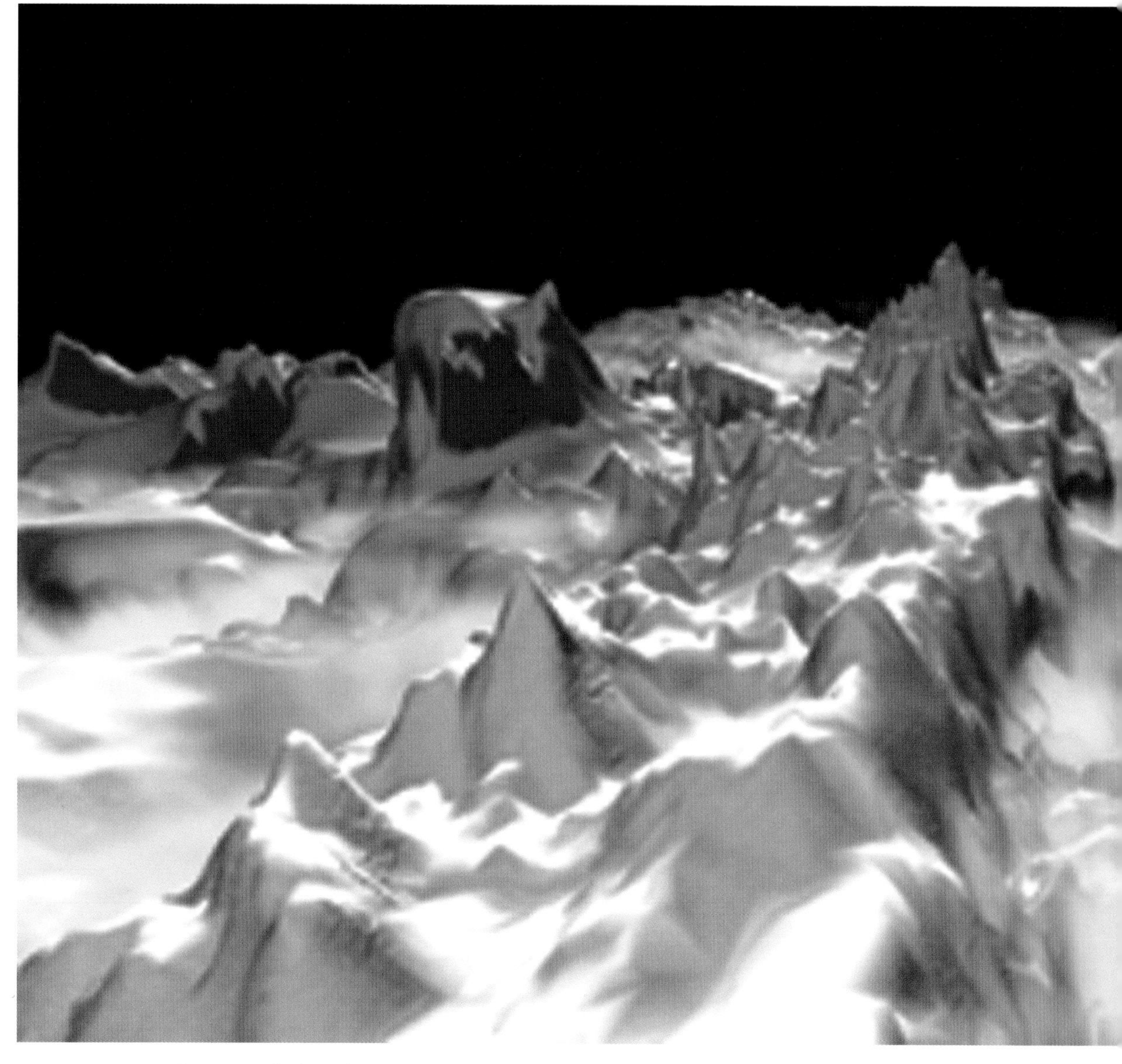

A planet too far from its star, or without a decent atmosphere to keep in the heat, will probably be too cold for life to get a start.

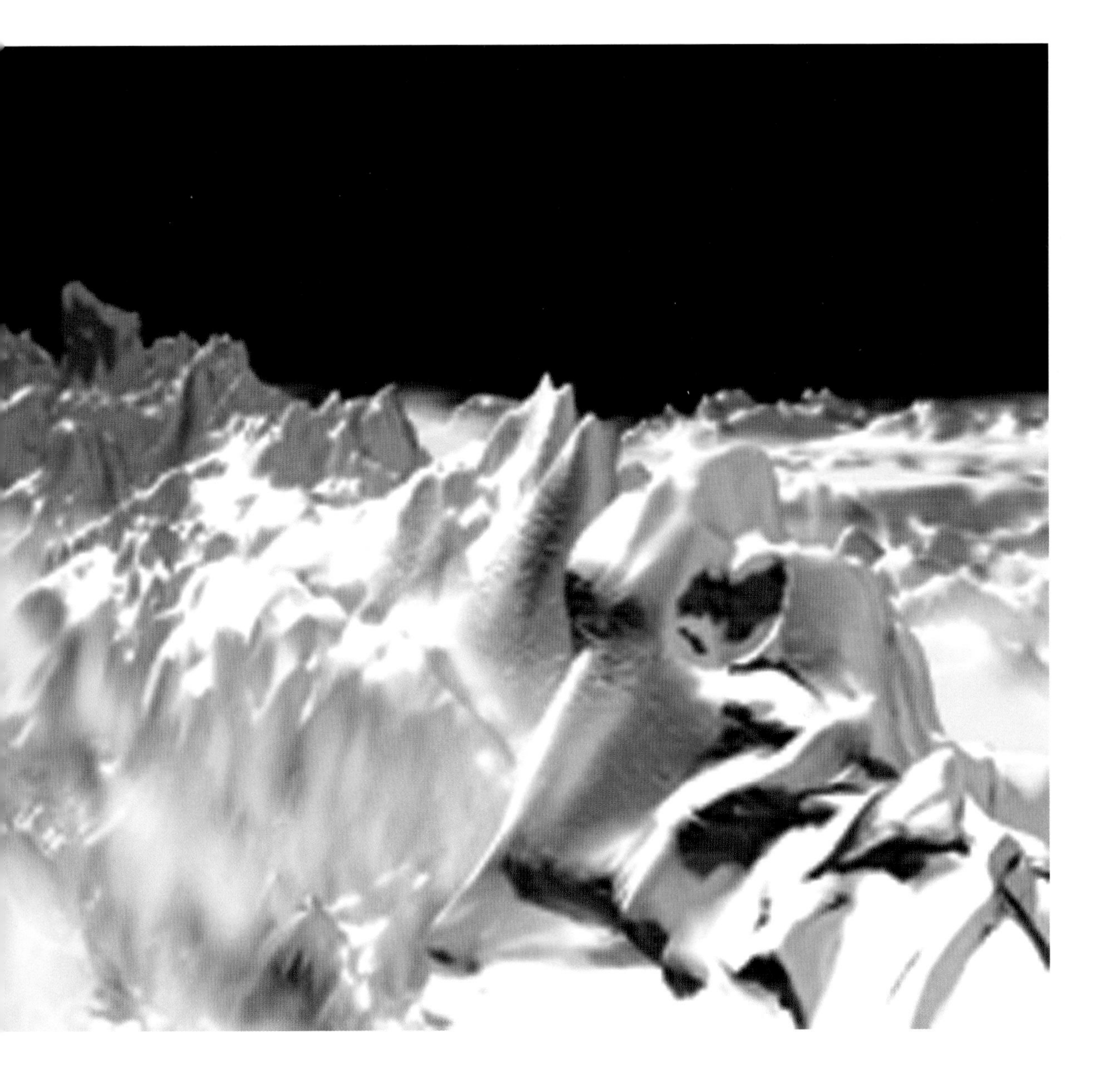

A planet too close to its star or covered with a thick, blanket-like atmosphere could be too hot for life to survive.

The bottom line is rather simple: although Sun-like stars seem the best bets for hosting planets habitable by intelligent beings, the overwhelming majority of stars could, in principle, be shining on worlds with life.

Bountiful planets

Good suns with planets in the habitable zone are great. But that's not enough. How many planets actually *have* liquid water? A quick reconnaissance of our own Solar System reveals rather few. To begin with, there seem to be two kinds of planets: those that are small and rocky, and those that are large and gassy. We live on a rocky world, a description that also applies to Mercury, Venus, Mars, and Pluto. All the other planets of our Solar System are gas giants, like Jupiter. They have deep atmospheres filled with methane and ammonia, and no solid surfaces on which life could grab a foothold. There are good reasons to think that planets around other stars would also fall into these two categories. But clearly, it's the rocky planets that are the more appealing habitats for life. It's true that many – if not most – of these small, solid worlds are undoubtedly barren balls of rock, without atmospheres, oceans, or any of the other encouraging features of our own planet. Even so, some of them – by chance – are likely to be rather similar to Earth.

The family of the Sun, posed for portraiture.

Mars is a perennially favorite location to hunt for simple life.

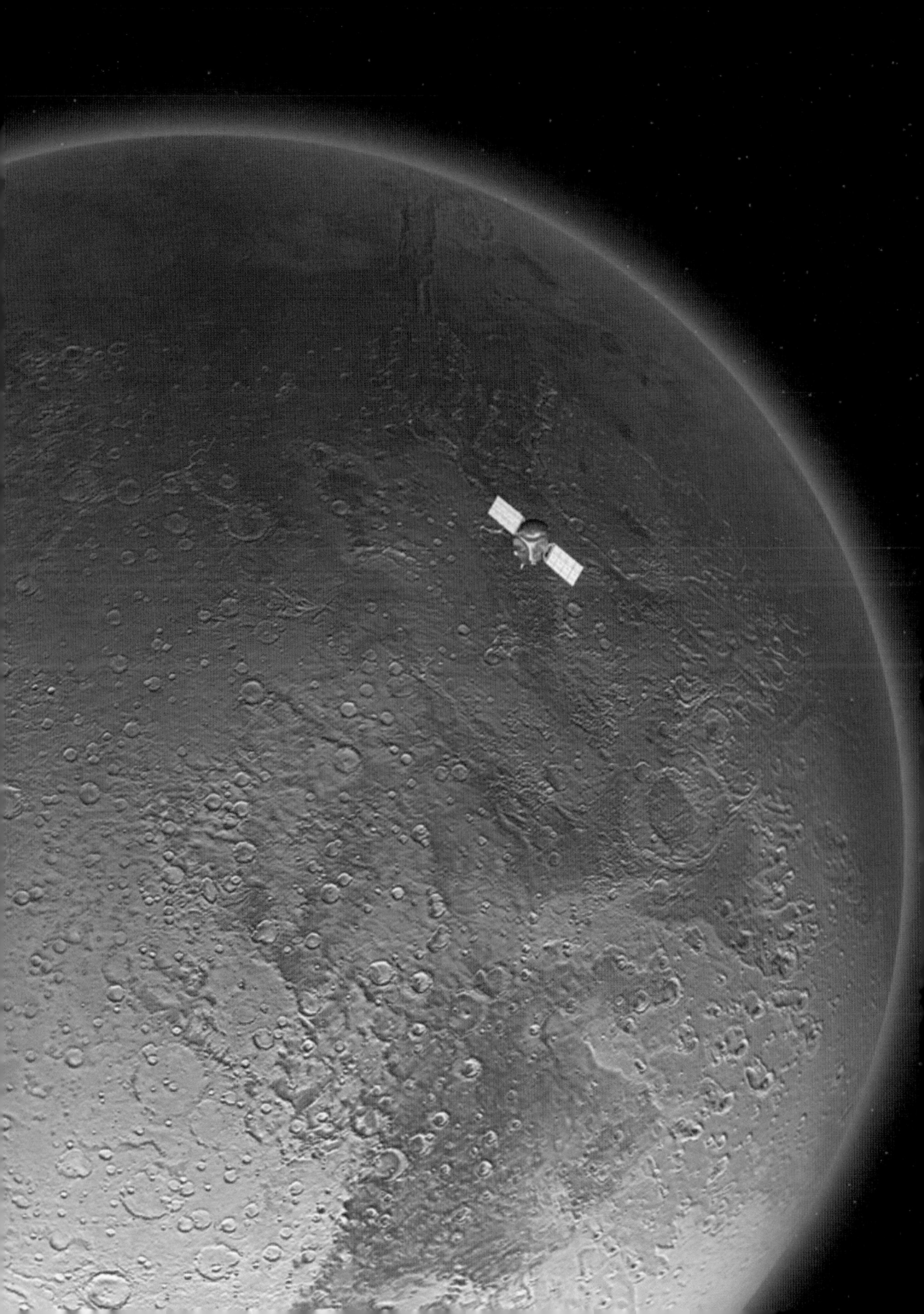

How many would that be? Again, we refer to our own Solar System, where at least three of the five rocky planets are dismal and dead. Mercury has no atmosphere and no liquid water to create life-giving seas. It's about as appealing a place for life as the Moon, which is to say, thoroughly unappealing. Venus is wrapped in a heavy, choking atmosphere a 100 times as thick as Earth's. The result is that the greenhouse effect has run amuck on Venus: the heat from the Sun has built to the point where daytime temperatures on this world are 500 °C, a bit too feverish for life. Distant Pluto, on the other hand, is frozen stiff.

But Earth has biology, and maybe Mars did too. The Red Planet seems to be a dead planet now, covered by a thin layer of cold air, and with no liquid water pooled or puddled on its surface. However, earlier in its history Mars may have had a thicker atmosphere, large oceans, and possibly some sort of life. In 1996, a handful of scientists claimed to have found evidence for microbe-size, fossilized Martians in a meteorite that was kicked off the Red Planet millions of years ago. If true, that would be astounding news,

(cont. on p. 30)

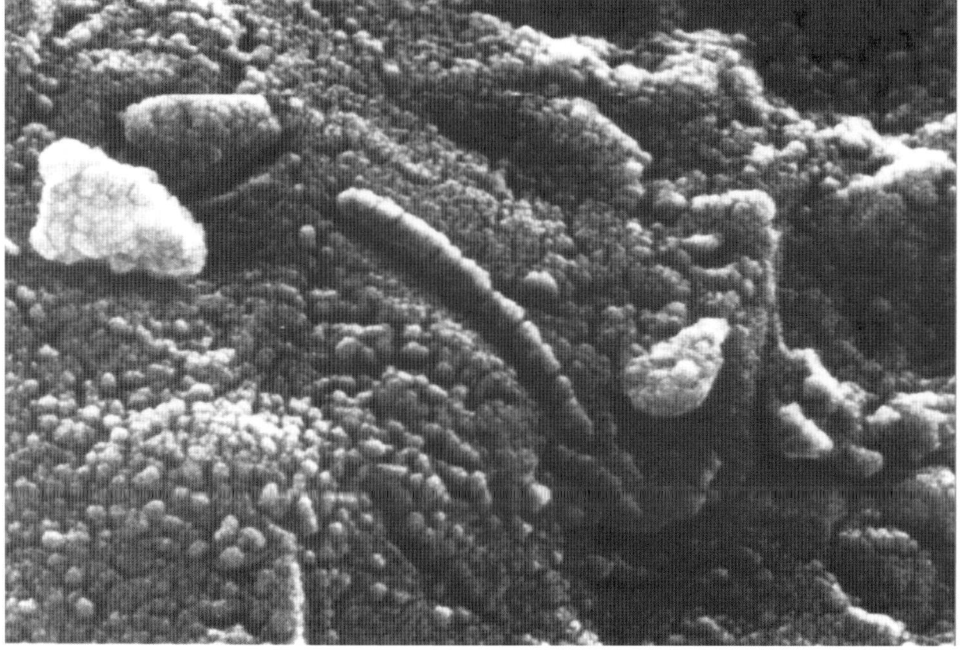

This tiny tube-like structure could be a fossilised Martian from 3.6 billion years ago.

Maybe Jupiter's major moons could have liquid oceans under their icy crusts.

Looking for simple life

SETI researchers think they know how to find and recognize intelligent life. A radio signal or flashing laser lights from other star systems are signals so clearly non-natural that they would be a sure indication of technologically sophisticated beings.

But what about looking for life that isn't so brainy? After all, several of the planets and moons of our own Solar System are suspected habitats for biology – either now or in the past – but no one believes that these places will harbor *intelligent* life. A reconnaissance for simple life on such nearby worlds will require more than just peering at them with our telescopes. A better approach – at least for Mars – is to launch sample return missions to collect bits of rock from the Red Planet's crispy, cold surface. But that type of "bring 'em back alive" (or dead) mission is at least a decade into the future. A less costly scheme is to send robot explorers. Needless to say, such roving robots will have to be clever enough to hunt for life in a barely understood, hostile environment. And since they might trip across life that's "not as we know it," they need to be flexible in their search techniques.

One place where unusual life may exist is near Saturn. In July, 2004, the US–European Cassini spacecraft will reach this giant ring world. Six months after arrival, it will swing by Saturn's moon Titan and release a probe named after the Dutch astronomer who first saw the moon in 1655, Christiaan Huygens. Titan is a bitterly cold place that scientists believe is pockmarked with lakes of liquid natural gas (the same compound used to heat homes). The action of weak sunlight on its atmosphere has fortified Titan with an abundance of complex, organic compounds. Could these have somehow led to biology? It may strain your brain to imagine that life could exist on a world where the temperatures hover at –180 °C on a warm day, but we won't know until we look.

And that's exactly what's going to happen when the Huygens probe – which weighs about as much as a pony – is released into Titan's smoggy atmosphere. The probe will enjoy a two-hour parachute drop before touching down on the surface. All this time it will be sampling the air (which is thicker than our own!) and sending back data on its temperature, composition, and even the speed of the winds. If Huygens survives the landing, it will have less than a half-hour to scout out the murky landscape before its batteries die. Unlike the Galileo probe dropped into Jupiter's atmosphere in 1995, this robotic explorer is outfitted with

The Cassini–Huygens probe parachutes onto Titan's intriguing surface.

cameras for making detailed photos of its surroundings. Even if Huygens settles down on to a liquid gas lake, it will still complete its mission, dutifully reporting the size of the waves as it floats in a strange ocean on a strange world.

Certainly one of the most enticing places to look for simple life nearby is Jupiter's moon, Europa. The challenge for the scientists hunting for "Europans" is that the granite-hard ice that covers the moon's oceans (assuming there *are* oceans) is 10–15 km thick. Many clever designs for landers that could somehow melt holes through this frozen skin have been proposed, but at the moment there are still no definite plans for putting a spacecraft on to the ice. An orbital

The Mars Exploration Rovers are each the size of a small car, but considerably more expensive.

reconnaissance mission that would use radar and other techniques to prove that the ocean exists, and could look for promising thin spots in the ice, has been planned, but probably won't be launched any earlier than 2009.

The most obvious place to look for life in the Solar System is, of course, Mars. NASA's Odyssey spacecraft, which has been orbiting the planet since early in 2002, has detected the presence of water in much of the Martian landscape down to a depth of at least one meter (the instruments couldn't probe any deeper than that). Presumably, most of these liquid assets are unappealing permafrost, but there's bound to be some depth at which the water is liquid, and life could thrive. In 2001, the Mars Global Surveyor snapped photos of gullies running down Martian hillsides, suggesting that pockets of liquid water might be hiding only a hundred meters or so below the surface.

This year (2003), NASA plans to launch two more rovers to explore Mars' rusty landscape, while the British are launching a Mars explorer called Beagle 2, as

The Beagle 2 Lander, carried by the ESA Mars Express Mission, is equipped to look for life.

part of the European Space Agency's Mars Express Mission. Beagle 2 has a mass spectrometer on board, an instrument that can analyze materials in a search for the compounds of life. The explorer will scoop up chunks of Martian material, and then incinerate them in the presence of oxygen. Any organic (carbon-containing) compounds in the chunks will form carbon dioxide. But is this gas due to biological compounds or not? Sorting out the biological from the non-biological carbon can be accomplished by burning the samples at different temperatures. For instance, most of the carbon compounds of your body will burn at a few hundred degrees Celsius. Diamond (which is pure carbon, and not alive), burns at much higher temperatures. The nice thing about this experiment is that it doesn't matter what any possible Martians' exact chemical construction is – if they're "carbon-based life forms," Beagle 2's mass spectrometer can detect them.

In addition Beagle 2 will sniff the Martian air for methane, a bacterial exhaust gas that could be wafting around the planet if life exists somewhere near its surface, even if far from the landing site. The rover will also have cameras aboard, and even a microscope.

Although no one is yet sure whether small, relatively simple life is lurking somewhere on the moons or planets of our Solar System, the intelligent life on Planet Number Three is doing its best to think up creative ways to track it down.

because it would be our first clue that life can be found on other worlds – indeed, found on the next world out from the Sun! If we could prove that life once existed (or perhaps still exists) on Mars, we would have a good reason to believe that many planets have life. However, the meteorite evidence is highly controversial, and the experts are still battling one another about whether the claim that life once existed on Mars should be believed or not.

But Mars is not the end of the story in our Solar System. To the initial surprise of staid astronomers everywhere, space probes have strongly suggested that several of Jupiter's large moons may actually have huge oceans of salty, liquid water hiding beneath their frozen outer skins. It's possible that, in the course of billions of years, life has appeared in these secluded seas, although it's unlikely to be either complex or intelligent. Beyond Jupiter, we have reason to think that Saturn's biggest moon, Titan, may have lakes of liquefied natural gas on its surface. Could it be that Titan has managed to produce some sort of life that's able to thrive in a dark, bitterly cold environment? The tantalizing thought of simple creatures living on the moons of the outer Solar System has greatly enlarged our concept of what is a "habitable zone."

Very crudely then, we see that around the Sun one-third of the Earth-size planets and several moons have, or may have had, life, and one in five of the planets developed intelligent beings. We have no idea if this is typical, but if it is, then we might expect that many millions of worlds in the Milky Way have produced thinking aliens, assuming that rocky planets are commonplace.

Do we actually *know* that rocky planets are commonplace? The answer is no, we don't. The reason is that small planets are hard to find. Indeed, *any* planets around other stars are hard to find! You can't just point a telescope at a nearby star and hope to see planets. Such small worlds – which only shine with the dim glow of reflected light – are at least a thousand million times fainter than the stars they are orbiting. Trying to see them in a telescope would be like trying to observe a moth around a Hollywood searchlight from 10,000 km away.

Despite the difficulty, astronomers have managed to find evidence that such planets exist. They have discovered more than a hundred planets around other stars since 1995 (the tally ratchets up at the rate of about

(*cont. on p. 34*)

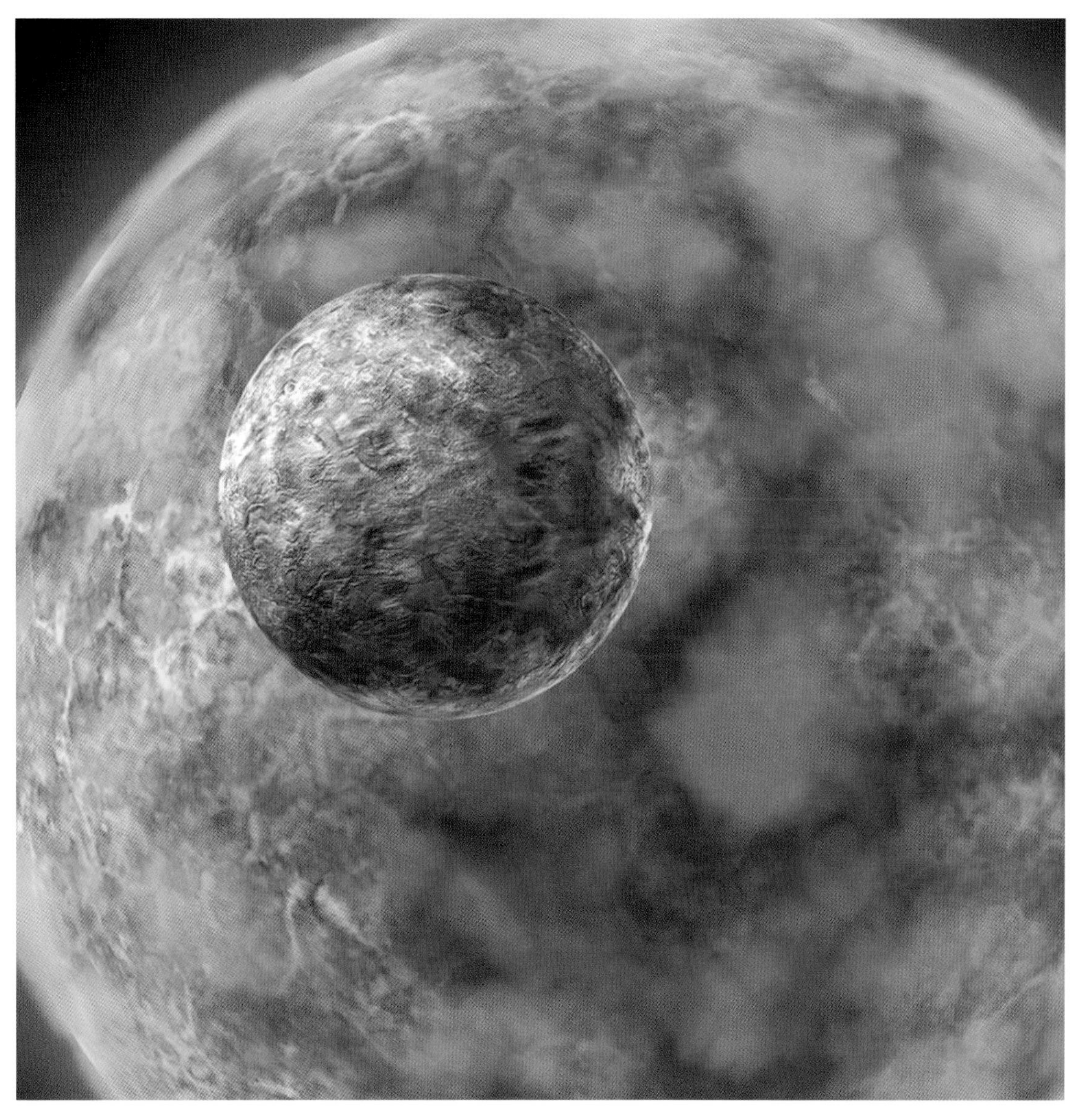

Many of the planets that have been found around other stars hug their parent star so closely that they are probably too hot for life.

Looking for Earth-size worlds

The first planet found around another ordinary sun orbits the star 51 Pegasi. This Jupiter-size world is close-in and exceedingly warm.

The Kepler Mission, a space telescope on the lookout for planets around other stars.

In recent years, astronomers have found that planets are common: at least 10 percent of Sun-like stars are accompanied by worlds that are roughly the size of Jupiter.

That's good news, of course. But it's unlikely that giant planets – with their deep, toxic atmospheres – are homes to life. Not surprisingly, the holy grail of planet detection is to find Earth-size worlds and, better still, to find those that are in the "habitable zone" of their star – at distances that would allow oceans of water to exist. There are several upcoming space programs that will press the search for such attractive worlds.

The first of these is a new NASA space-borne telescope, called the Kepler Mission. Kepler's strategy is quite straightforward: to stare at the stars. As astronomy buffs are well aware, every few years we see Mercury or Venus cross the Sun's disk. During these transits, the planet blocks some of the light from the Sun – not much, mind you; typically about 0.01 percent for Venus and even less for diminutive Mercury. The Kepler Mission, which consists of a medium-size space telescope, will look for such transits around *other* stars. The amount of blocked light, and the length of the transit (typically many hours), will clue astronomers to both the size and orbit of any detected worlds. Of course, to see a transit requires two things: (1) the orbit of the planet needs to be correctly lined up with the viewer, and (2) the telescope needs to be pointing at the star when the transit occurs. Since we don't know anything about the sought-for planets' orbits, the researchers are arranging for the Kepler telescope to stake out a whole slew of stars, and monitor their individual brightness continuously for four years. In addition, careful attention to design means that brightness variations smaller than 20 parts per million will be noted by Kepler. Earth-bound telescopes – which must peer through our churning atmosphere – could never achieve such accuracy.

The Eddington Mission, another planet-hunting space telescope.

Kepler will be blasted into orbit sometime in 2007, to follow Earth around the Sun and keep its well-calibrated eye on a chunk of sky that's roughly the size of two Big Dippers. It will stare at 100,000 stars. What will Kepler find? That's the exciting part, because we don't know. If Earth-size planets are common, the mission will turn up about 50 of them in orbits comparable to our own. Thousands of closer-in and/or larger planets could be found. Even bulky moons around Jupiter-size planets might block enough starlight to make their presence known. All such worlds are potential abodes for life.

The European Space Agency is also launching a mission called Eddington that will have the capability to look for planets. Scheduled for blast off in 2008 and traveling to an orbit well away from the Earth, Eddington will also carry a sensitive light detector. Eddington is primarily intended to look for "starquakes," stellar hiccups that can tell astronomers a great deal about what the stars are made of. But like Kepler, it will also note if any planets transit in front of the star.

These missions are both dramatic and important. As Jon Jenkins, one of the Kepler researchers, said: "The discovery of the New World was accomplished in a few decades, and now the discovery of entirely new worlds around other stars will also take place in the same length of time. This is something that will happen only once in human history."

How common are Earth-size planets? These new orbiting telescopes will finally answer this incredibly interesting question.

one new world every three weeks), and this is an encouraging sign for those of us hoping to find life elsewhere.[2] Indeed, more than one in ten of the stars studied by the astronomers has planets. But all of these worlds have been found by measuring the very slight wobbles that such planets cause in their host stars, so we don't actually see the planets themselves; we merely measure their wobbling suns. Regrettably, small Earth-size worlds only produce small wobbles, and these cannot be picked out with today's telescopes. So it's not surprising that virtually all of the planets discovered orbiting other stars are giant, economy size – at least the heft of Jupiter. If we could actually see these heavyweight worlds, we'd be likely to discover that they are gassy and unappealing for life (although their moons might be OK). But since we have a mix of big, gassy worlds and small, rocky planets in our own Solar System, we expect that at least some of the large worlds around other stars are also accompanied by small, solid planets. Proof one way or the other will only come when we turn new telescopes towards the heavens that are capable of discovering these hard-to-see, Earth-size worlds.

Are they out there?

As we've seen, the requirements for life seem to be widely distributed, and it's tempting to think that the Cosmos is chock-a-block with life, and even thinking beings. We've noted that our telescopes can catch the light from as many stars as there are grains of sand in the Sahara. How odd would it be, and how remarkable, if our Sun were the only grain of sand where intelligent beings have gained a foothold. To quote Ellie Arroway in the movie *Contact*, that would be a serious "waste of space."

But is that just wishful thinking? Ellie Arroway may care about wasted space, but does Nature? It's conceivable that the entire Universe, or at least a large part of it, is occupied only by us, and the rest *is* wasted. It's possible that the massive nebulae and star fields of the Milky Way exist for no greater purpose than to be photographed, framed, and hung on the office walls of your local astronomer. We cannot yet rule out the possibility that humans are the only life forms able to look at the worlds around them with understanding. Of course, this would mean that *Homo sapiens* is exceedingly special, and extraordinarily lonely.

[2] A good website for keeping up with the discovery of new planets is http://planetquest.jpl.nasa.gov/.

However, if you think about it, you'll realize that there's no way we can prove we're alone. There's no experiment we can do that will conclusively show that no intelligent beings are out there, because the aliens could always evade our telescopes. We can't look everywhere all the time. But what's new and amazing is that now, in the twenty-first century, we can prove we're *not* alone, simply by searching for convincing proof of other beings, from other worlds. And as we'll describe later in this book, the search is on.

CHAPTER 2

What might the aliens be like?

The first thing we'd probably want to know about any extraterrestrials we discover is what they look like. In the movies, the good aliens often resemble us. They have small, hairless bodies, big heads, and appealing, almond-shaped eyes. That makes sense for a film, since extraterrestrials resembling 2-ton, sightless turtles or something equally odd would be very hard to relate to, and you wouldn't care much about them one way or the other.

Nature doesn't necessarily share Hollywood's preference for humanoid aliens. Evolution has provided an enormous range of creatures on Earth, and presumably would do so on any other planet. So at a first glance it seems as if there's really no guessing Jo Alien's appearance until we either meet him,[1] or receive a photograph sent electronically from his home planet. But this is not entirely true. On the basis of what we know of life on Earth, as well as the simple laws of physics, we can make a few reasonable assumptions about life elsewhere. At least, we can do this if we have some idea of what Jo's home planet is like. That involves asking a few questions about the start of life, and the conditions necessary to keep it going.

The beginnings of biology

How and where would life get underway on an alien planet? We can gain some insight into this by considering how this happened on Earth. Unfortunately, life's early days on our planet are hard to trace. This is because the evidence is buried in rocks. More particularly, it is buried in old *sedimentary* rocks that formed at the bottom of the ocean. It is there, on the sea floor, that the remains of the earliest organisms accumulated,

[1] Throughout this book we refer to Jo Alien as "him" for convenience and at the whim of one of the authors. We don't mean to imply anything about alien genders, however.

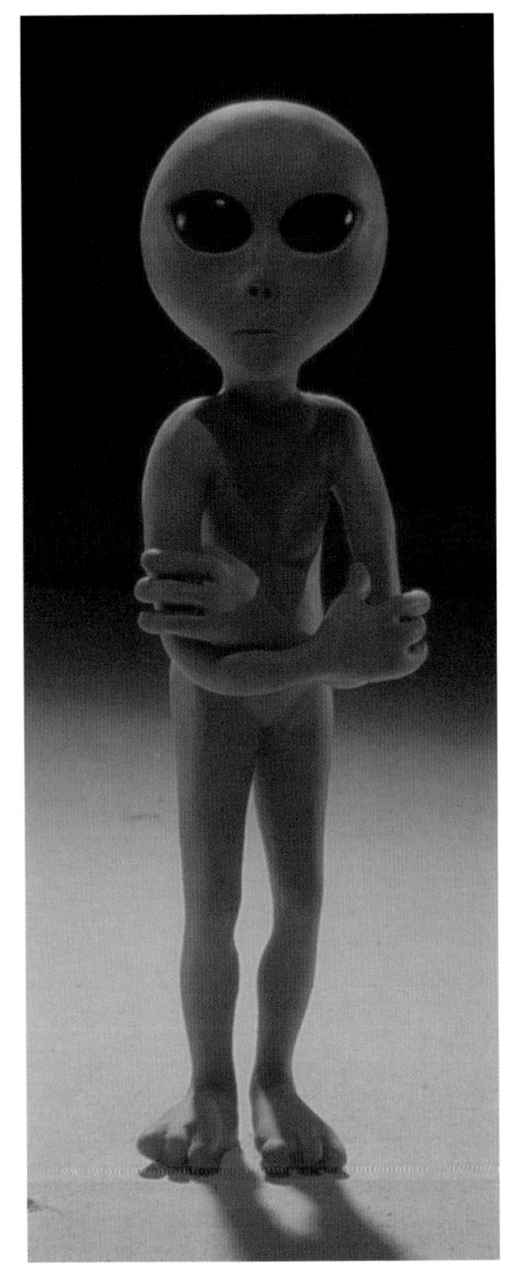

A typical – and rather humanoid – Hollywood alien. Real aliens won't look like this.

and were occasionally interred and preserved in the mud before they could decay. These sedimentary rock beds hardened and became fossil layer-cakes for the delight of today's paleontologists. Unfortunately, the most ancient and, from the standpoint of learning something of life's early history, most interesting sedimentary rocks – those older than about 3.5 billion years – are badly messed up; tortured and twisted during Earth's tumultuous youth. Nonetheless, we have found what appear to be fossilized microbes that date back this far, and even some tentative evidence for life in rocks as old as 3.8 billion years.

That's an interesting result. We know with certainty that the Earth (and all the other planets) were born together with the Sun 4.6 billion years ago. But we also know that for their first half-billion years, the planets were unwitting targets in a giant shooting gallery. Countless hunks of rock, some thousands of kilometers in size, were hurtling around the young Solar System. These were the leftover building blocks of the planets. They routinely cannoned into our world, pummeling its surface. To get some idea of how dramatic and hellish these early years were, simply look at the Moon. Most of the lunar landscape is cluttered with craters, the scars caused by rocks that mercilessly bombarded the Moon about 4 billion years ago. These bear silent witness to the Solar System's violent early years. But if you check out the large dark blotches on the Moon, which seventeenth-century astronomers called "maria" (Latin for "seas") because they thought they were looking at

A snapshot of our Solar System forming, 4.6 billion years ago.

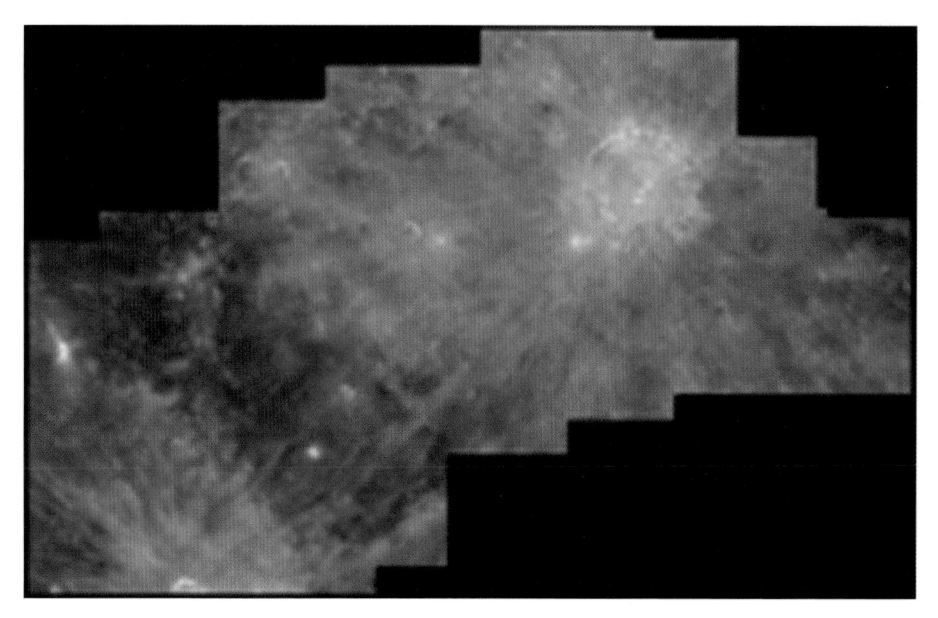

This Hubble telescope image shows how our Moon has been scarred by the bombardment of countless chunks of rock.

On Earth, water has been the cradle of life. On other worlds, this will probably be the case as well.

liquid oceans, you'll note that these areas are smooth and nearly crater-free. By examining Moon rocks brought back by astronauts, we know that the maria are not quite so old – dating back less than about 3.8 billion years.

In other words, the Moon provides us with good evidence that by 3.8 billion years ago, the lethal pounding of our planet had let up. And since the oldest known fossils are between 3.5 and 3.8 billion years old, this tells us that nearly as soon as life *could* survive on our planet, there *was* life (at least within a few hundred million years, which for geologists isn't much). Since Nature didn't take an enormous amount of time to get life underway on Earth, this strongly suggests that – whatever the details – the project wasn't that difficult.

Difficult or not, the precise way in which life began is still a mystery. In fact, we don't even know where it first sprang up. Charles Darwin, in a private letter written in 1871, suggested that the initial stirrings of life may have taken place "in some warm little pond." This general idea makes sense. As we've already noted, liquid water is a wonderful medium for encouraging the chemical processes of life. In addition, all chemical reactions are speeded by heat, a fact that may have motivated Darwin's preference for a "warm" pond.

Something that Darwin didn't know is that anything camped out on the early Earth's surface would have been exposed to the Sun's brutal ultraviolet light. In those distant days, there was very little oxygen in the air. (Oxygen arose only a few billion years later, as microbes busy with photosynthesis slowly filled the atmosphere with their exhaust gases, including oxygen.) Consequently, there was no atmospheric ozone – a form of oxygen – to screen out most of the ultraviolet and protect delicate life on the rocks below from fatal sunburn. This is a reason to suspect that life began in a sheltered environment; for example, underwater.

In recent years, biologists have discovered that life can thrive on the ocean floor, miles beneath the surface, in waters that are darker than a cave, and at temperatures that are hotter than steaming coffee. In these hidden ecological niches, water brought to a boil by the Earth's internal heat comes bubbling up through small cracks, or vents, on the ocean bottom. It doesn't sound like a great place to hang out, but life exists there, nonetheless. Indeed, such deep-sea vents may be the place where the earliest ingredients of life came together to build the first microbe. Rather than a "warm little pond," this would be a "large hot ocean" scenario for the start of life.

Whatever the truth, at some point, unnoticed and unrecorded, complex molecules appeared that were able to produce copies of themselves. Once this occurred, natural selection would favor those that could replicate quickly and with little error. The first steps down the yellow brick road of evolution were taken, and the rest of the journey would be, relatively speaking, just a long walk leading to microbes, multicellular life, and (eventually) humankind.

We believe, as we've noted, that life on Earth began in water. This seems so reasonable and right that we strongly suspect that the same is true for most alien biology. A watery past has influenced our appearance and bodily equipment, and it would possibly affect the aliens' body plan as well. But before considering how a wet early lifestyle might shape Jo Alien, we need to inquire about yet another aspect of existence: where do living things get the fuel they need to survive?

What could power life?

Life requires an energy source. This may not be so obvious when you sprawl on the couch and do nothing more than occasionally punch the TV remote. But even in this low-energy state, your body is still putting out as much heat as a 75-watt bulb (although not as much light). Where does the energy for life come from?

It depends on what type of life you're discussing. We've noted that there are living things huddled around the seething, hot water spewed from deep-sea vents. This is water that has been brought to scorching temperatures by Earth's internal heat – the same heat that moves the continents and powers volcanoes. But what's the source of this planetary body heat? To begin with, the Earth's interior is still warm from the countless collisions that produced our planet in the first place. There's also energy that was gained during the formation of Earth's metal core. When the planet was still hot and soft, heavier elements (iron and nickel, mostly) sank to the center, releasing energy as they fell. Finally, there's an on-going heat source: the slow decay of radioactive materials in our planet's interior. All

The ingredients of life

Most biologists agree that the first step in getting life started is to assemble the necessary ingredients. These consist of simple molecules containing carbon since, as every *Star Trek* fan knows, life on Earth is carbon-based. This is not an accident. Carbon is the fourth most abundant element in the Universe, so just about every world will have a lot of this important material. But the real attraction of carbon is that it is able to easily hook up with other elements (particularly hydrogen, oxygen, and nitrogen) to form an enormous variety of complex molecules useful for life. No other element does this as well as carbon, so it's more likely than not that extraterrestrial life will also be carbon-based.

The most important *molecular* building blocks for terrestrial biology are small, nitrogen-bearing compounds known as amino acids. Twenty different amino acids, sporting names like leucine, lysine, and valine, are found in earthly life (this is far smaller than the number of *possible* amino acids, incidentally). Chains of amino acids make up proteins, the essential components of living cells.

Carbon is made by stars. But where did the first amino acids come from? In the 1950s, two biochemists at the University of Chicago performed a simple experiment that suggested an answer. They filled a laboratory flask with methane, ammonia, and water. These, they thought, were the major chemical compounds in the Earth's atmosphere 4 billion years ago. Then they zapped this mixture with electric sparks for a week (simulating the effects of lightning). The contents of the flask turned into a brown goo, made up of lots of organic molecules, including amino acids. They concluded that simple chemistry, energized by lightning or the Sun, could have produced amino acids in the Earth's ancient atmosphere.

As it turns out, we now believe that 4 billion years ago the atmosphere actually contained rather little methane and ammonia, but was dominated by carbon dioxide. Nonetheless, variations of the Chicago experiment still work: it's possible to pump energy into a more realistic mixture of gases and have them naturally cook up the basic ingredients of biology. If this actually happened in the atmosphere of the young Earth, then the building blocks of life would have incessantly rained down into the oceans or possibly some warm little ponds.

It seems plausible that biology's essential molecules could have been home grown. But there are other points of view. Meteors, asteroids, and comets also contain amino acids that were apparently produced in the liquid water

The early development of life on Earth is often described as taking place in a primordial soup.

occasionally found in such space rocks. Recent experiments have shown that even interstellar dust grains – bits of icy material that are found in space everywhere that stars and planets form – can also provoke the chemical reactions that make amino acids. Since dust, meteors, comets, and asteroids have been falling to Earth since it was formed, perhaps the raw materials of life were originally produced in the cool, dark recesses of the Solar System and then – by occasional accident – deposited on to our planet.

An extension of this intriguing possibility is the idea that life itself, and not just its building blocks, came to us via space rocks – not those that careen around our Solar System, but chunks of material that managed to make a journey from another star system altogether. Could it be that a distant planet of which we are totally unaware is the source of life on our planet? Is it possible that a rock knocked off such a far-off world by an impact contained living spores that managed to survive a trip of 100,000 years or more to

Did life get "seeded" on Earth by passing comets?

eventually infect a young Earth? This idea, which has the sexy name "panspermia," is beguiling – and controversial. We have no proof that it's true, but we also haven't ruled it out.

of these heat mechanisms could be present on other worlds, and might provide energy for life in underwater environments.

But that's not all. There's another possible source of calories for biology known as tidal heating. This odd phenomenon could be important in keeping water warm in multiple-moon systems, such as those that orbit Jupiter. Four of Jupiter's many moons are big, approximately the size of our own Moon or larger, and close-in to the planet. Like close-in moons everywhere, they would like nothing better than to settle into nice, more-or-less circular orbits, with one side always turned towards Jupiter, in the same way that one side of the Moon always faces Earth. Unfortunately, the moons can never achieve this state of satellite bliss. That's because they tug on one another, and the inner moons in particular are nudged into egg-shaped orbits.

Io is the most volcanically active place in the Solar System. It is kept warm by the continual squeezing and stretching from Jupiter's gravity.

The result is that Jupiter's strong gravitational pull on these moons is ever-changing as their orbits swing them closer to, then farther from, the giant planet. This stretches and squeezes the hapless moons, much the way that you might stretch bread dough. The dough gets warm due to the friction of stretching, and so, too, do Jupiter's moons. This heat can keep water near the surface in a liquid state – water that would otherwise be solid ice. Europa, Ganymede, and Callisto might have enormous, dark oceans permanently hidden from our view by a thick ice crust. Tidal heating could not only serve to keep such oceans from freezing, but would also provide hot water to bubble up from the sea floor – energy to fuel the processes of life.

There are, undoubtedly, a wide variety of possible energy sources for life. But the really interesting biology – at least on our planet – feeds off an obvious and abundant power supply: sunlight. The calories necessary to keep you warm, to flex your muscles, and to fire your imagination are all derived from sunlight. This energy is captured by the well-known process of plant photosynthesis, which is the first link in the food chain. Green plants use sunlight to convert water and carbon dioxide (two inorganic and highly abundant compounds) into glucose, a sugar that can be stored in the plant, or into other molecules such as sucrose, fructose, or starches. When you chow down on vegetables and fruits, it's more than just pleasing Mom: it's taking advantage of the sunlight stored by these plants. Of course, the same is true if you dine on steak, lobster, or fried ants, all of which have energy in their tissues from sunlight sources (either because they ate plants, or because they ate other animals that ate plants).

There are lots of plants here occupying their time by storing energy from sunlight.

Sunlight is available everywhere on the surface of the Earth, which is why – at least on our planet – it's biology's favorite fuel. As we've already discussed, life is most likely to appear on planets around stars that are either Sun-like, or perhaps a bit smaller. That guarantees enough sunlight to energize photosynthesis or something like it. So it's not at all improbable that the food chain on other worlds also begins with vegetables, fruits, and other starlight-powered life forms.

The influence of your planet

There are many possible environments that might spawn life. But while simple creatures could arise in places quite unlike the sunlit continents of Earth, there are good reasons to think that intelligent life won't. Life is most likely to have a watery beginning, but just because a world has

We believe that intelligent life is more likely to spring up on a planet quite like Earth. What you believe is up to you.

an ocean doesn't mean it could spawn a sophisticated, thinking Jo Alien. Deep-sea vents in inky-black oceans permanently capped by thick ice, such as might exist on Jupiter's moons, simply don't offer the abundant energy resources of planets where both land and sea are exposed to the light of their sun. If life exists in such ice-encrusted worlds, it will be low-grade and limited.

Simply on the basis of energy requirements, it seems that our best bet for finding intelligence is a rocky planet with a liquid, surface ocean, so that plants and animals floating in the top layers can take advantage of the sunlight. An atmosphere is handy for keeping the ocean from either freezing or evaporating away. It's also useful for protecting the life below from some of the dangerous ultraviolet light and high-speed particles from space that would otherwise be a constant menace. Finally, we expect that sophisticated aliens would be more likely to develop technology if they were land dwellers, for reasons we'll discuss below.

In other words, Jo Alien's planet will, in several important respects, resemble ours. What sort of complex life might such planets produce? What would it look like? As we address this question, we'll only consider animals. Why? One reason is that large plants tend to be fixed in place. Animals, on the other hand, are mobile. This is a big advantage in that being mobile means you can collect energy faster – energy you need if you sport a big brain. Consider: it might take weeks for Brussels sprouts to grow in the Sun. But as a mobile animal, you can quickly find and eat a handful of sprouts, and burn up weeks' worth of laboriously stored calories in a few minutes. That's what animals do. It's faster. No offense to plant lovers, but Jo Alien will be an animal of some kind.[2]

It's also likely that Jo will indulge in eating *other* animals, as meaty meals have highly concentrated reserves of both energy and the raw materials (proteins) of life. Dining on an occasional sprout isn't going to adequately fuel Jo's brain for long. Human brains consume about one-third of the total energy used by our body, even though they're only about one-fiftieth of the mass. We believe Jo Alien's brain will be an energy hog, too. He could get this energy by spending the entire day grazing on salads, rather like cows and sheep, but that wouldn't leave time for much else. Or he could take a few minutes to catch a fatty, protein-rich animal. Consequently, Jo will probably have at least a few sharp teeth, or some other cunning way of consuming animals.

While we anticipate that alien animals will run on stored energy, they won't all have the same appearance. A quick check of the local zoo will convince you that animals can be built with a wide variety of designs. Nonetheless, there are a few features that seem to be so commonplace, so useful, that we expect them to be found on any inhabitants of worlds where conditions approximate Earth.

To begin with, abundant sunlight is useful for giving detailed information about the environment, at least if you have eyes. This is such an obviously handy development that just about every type of moderately complex animal on Earth has some sort of eye. There are small, worm-like critters with light-sensitive cells on their surfaces. Diminutive insects and marine creatures have eyes (often quite a few). And of course back-boned creatures,

[2] Apologies to all the *Day of the Triffids* fans.

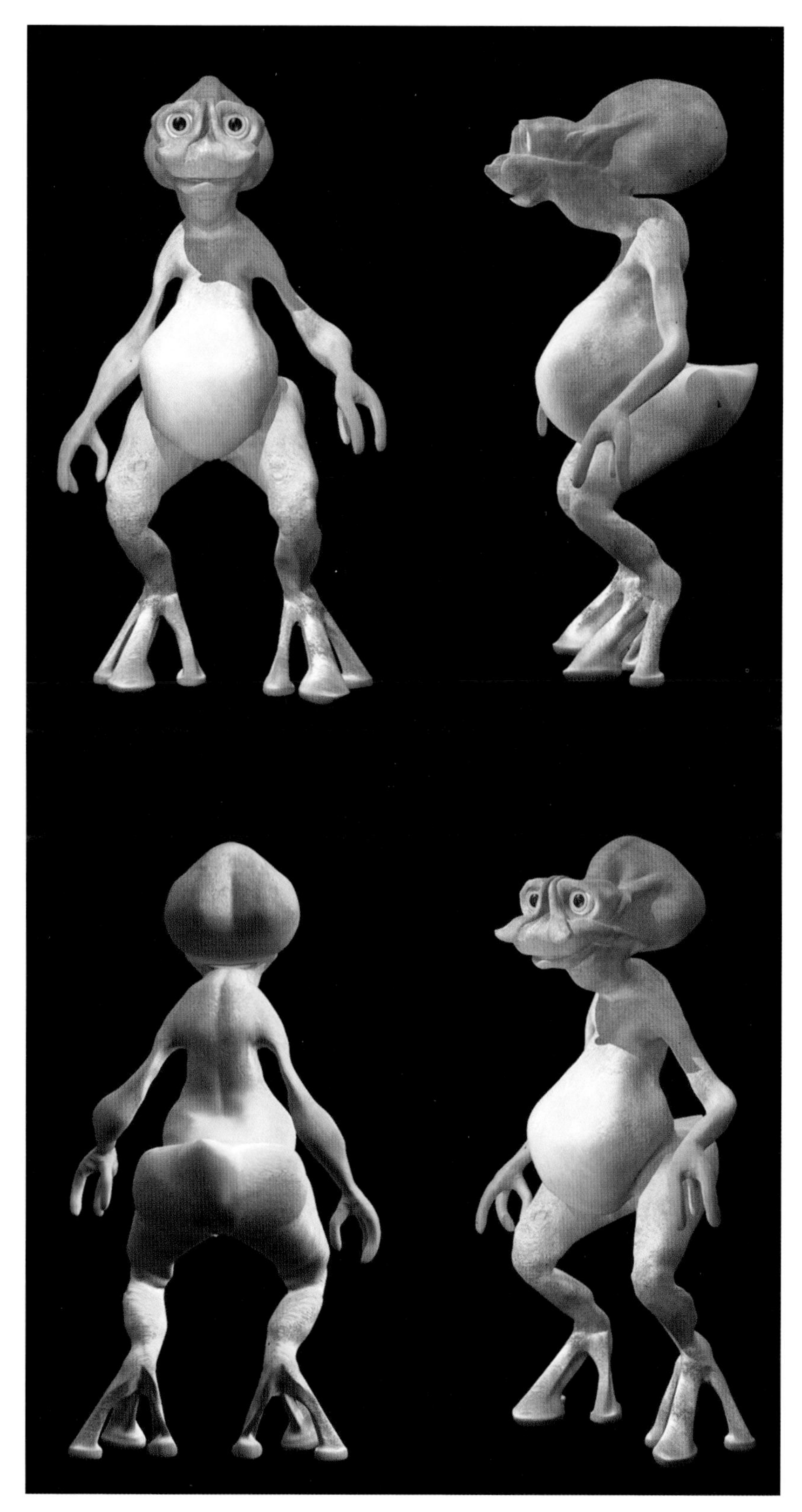

Meet Jo, a human-like imaginary alien.

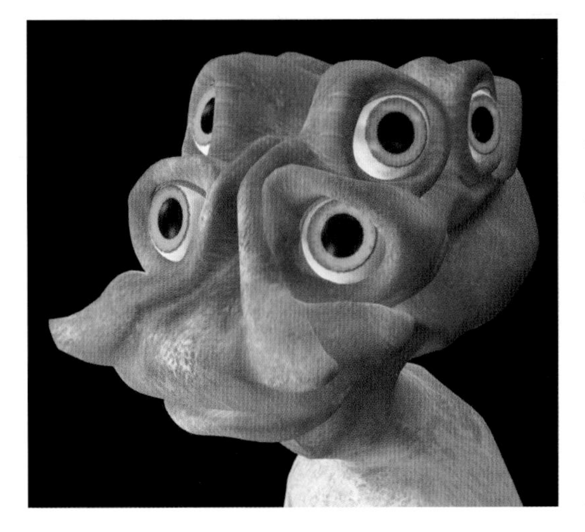
This many eyes would require too much brain for the gain.

such as mammals and birds, generally have sophisticated eyes with lenses to form images (much like a camera). Eyes are everywhere, and it's more than likely that Jo Alien will have them.

But how many eyes? There's a good reason to prefer two over one. With two eyes you get the benefit of 3-D vision, which helps to estimate the distance to the next tree branch (if you're a swinging tree-dweller) or to catch dinner (if you're trying to supply the main dish for a barbecue). To have 3-D vision requires that both eyes mostly see the same scene; that they both face in the same direction. That's why many earthly carnivores have forward-facing eyes, while preyed-upon creatures often find it better to have eyes on both sides of their heads so they can survey more of the landscape and escape if necessary. So we suspect that Jo Alien will have at least two, forward-facing eyes.

Could he have more? A lot of movie and TV aliens have three eyes, probably to make it easy to sort out the extraterrestrials from the terrestrials on the screen. Having three or more eyes isn't common on Earth, although some lizards boast a very poor third eye in the middle of their foreheads to keep track of how long they've sat in the Sun. One might think that a pair of image-forming eyes on the back of the head might help Jo Alien spot enemies, and this is certainly a possibility. However, more eyes means that more brain power (and more energy consumption) is necessary to handle the processing. We can't say that Jo will have only two eyes, but we can safely predict that he won't have dozens.

What capability will Jo's eyes have? Human eyes are sensitive to a rather narrow range of wavelengths (corresponding to the colors of the rainbow, the spectrum), but there are plenty of other wavelengths of light that are redder than red, and bluer than blue. There's lots of this light around, but we can't see it. The reason is historical. Eyes developed before animals had emerged from the oceans – the first eye designs were for creatures living underwater. It turns out that only a very limited range of light waves (or radio waves) actually penetrates water. Being able to see a wider range of colors might be useful on land, but our eyes were "locked in" to a more limited range a half-billion years ago, when our ancestors still dwelt in the sea.

Maybe Jo will be tuned into senses we don't have, such as being able to detect magnetic fields.

What about Jo's eyes? If life on his world could quickly colonize the continents, his eyes might not be as limited as ours. But life on Earth stayed underwater for billions of years for good reason. Until enough oxygen built up in our atmosphere to create a shield of ozone, it was dangerous for any undersea life to venture on to dry land. Something similar might have happened on Jo Alien's world, but it could also be the case that he evolved on a planet around a star smaller than the Sun – one that pumps out less of the dangerous ultraviolet light. In that case, Jo Alien's range of perceived colors could far outstrip our own.

A millipede seems to cope with lots of legs but its brain doesn't have to do much else.

What about ears and noses? Both are useful if you live in a medium – either water or air – that can conduct sound and convey scent molecules your way. Although they aren't quite as informative as being able to see (it's hard to determine the exact source of either a sound or a smell unless you're an animal blessed with truly gargantuan ears or noses), they can be *more* useful in the dark, or when forest vegetation or murky waters block the view. Again, since we expect that Jo Alien's world will have both liquid water and an atmosphere, smells and sounds are senses that make sense.

What about locomotion? On Earth, jointed appendages – we call them "legs" – are the favored technique for hauling oneself around the landscape (unless you're a snake). Even flying animals have legs. Is this just an

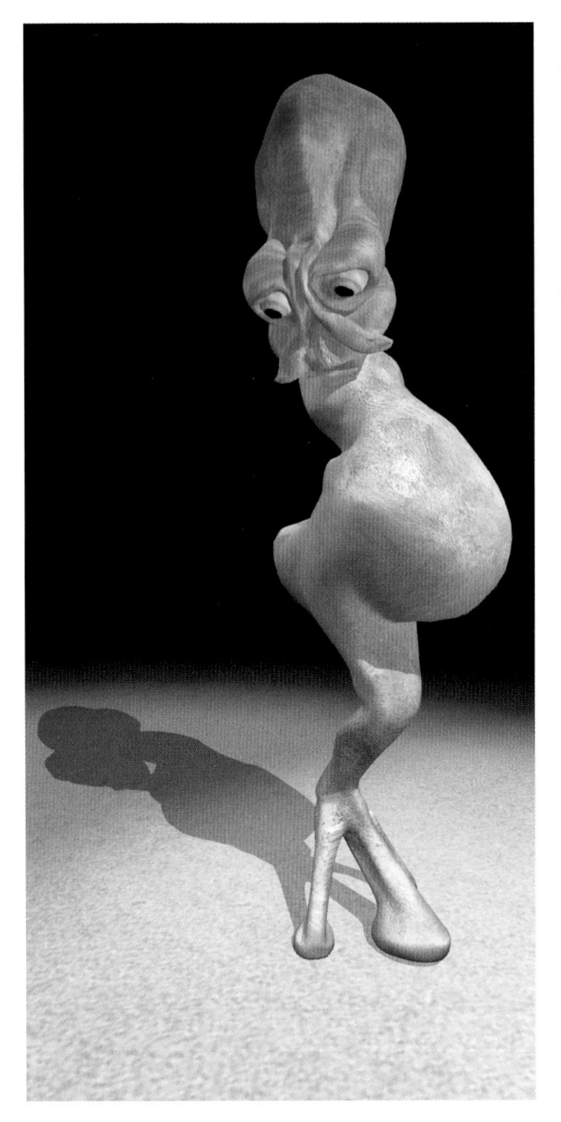

Having only one appendage makes getting around a bit awkward.

evolutionary accident? Might Jo Alien have wheels, rather than legs? After all, cars are faster than cheetahs.

Alas, wheels work best on prepared geography, which is to say, roads or rails, and planets aren't usually supplied with highways or railways as original equipment. Wheeled transport is difficult on rocky or sandy ground, although if Jo Alien had six or eight wheels – such as we build on to our Martian rovers – or if his wheels were economy size, like those fitted to monster trucks – he might be able to get around on tough terrain. But frankly, it's hard to imagine that wheels evolve as a major method of biological locomotion on anyone's world unless it's a very smooth one. Legs are a better bet. At least two. Maybe four. But having too many legs is like having too many eyes – your brain will be fried just handling it all. In addition, Jo will need a few appendages so that he can build things. Having just one appendage would make it very awkward to construct a rocket or a radio.

How big is Jo likely to be? There's little doubt that intelligence requires a large brain, since you need a lot of neurons (or their extraterrestrial equivalent) to think. The human brain weighs in at 1.4 kg, which is large compared to even the *total weight* of a lot of animals. We don't expect Jo Alien to be rat-sized or smaller unless evolution on his planet has managed to produce much smaller neurons than ours (something we can't rule out).

But is there an *upper* limit to the size of a thinking alien? There is if he lives on land. This is because, as you scale up a creature, you increase

If Jo is smaller than a rat, he might not have enough neurons to be smart.

its weight much faster than its strength. Double the height of an animal, and its muscles will have four times the cross-sectional area, and therefore boast four times the strength. But the creature itself will be twice as high, twice as wide, and twice as deep – in other words, eight times as heavy. So it is four times as strong and eight times as hefty, which explains why smaller animals on Earth can jump much higher than we can, compared to our size. Fleas are good examples, able to make a leap that is equivalent to a man jumping 300 meters into the air. (Among earthly hominids, only Superman can pull this off.) Ants can carry loads several times their own weight, but few of your neighbors can. However, if you scaled-up these insects a few thousand times so that they were the size of ponies, their weight would increase by many billions of times, and they would collapse in an unappetizing heap of cracked exoskeleton and internal goo. The next time you see scaled-up bugs threatening the populace at the local cinema, you can discreetly point out to other filmgoers that such monster insects simply don't stand up.

We expect that if Jo Alien's planet is roughly the size of Earth (and this makes sense for a rocky world that can have an atmosphere and oceans),

And a giant-size Jo wouldn't have the strength to support his own weight.

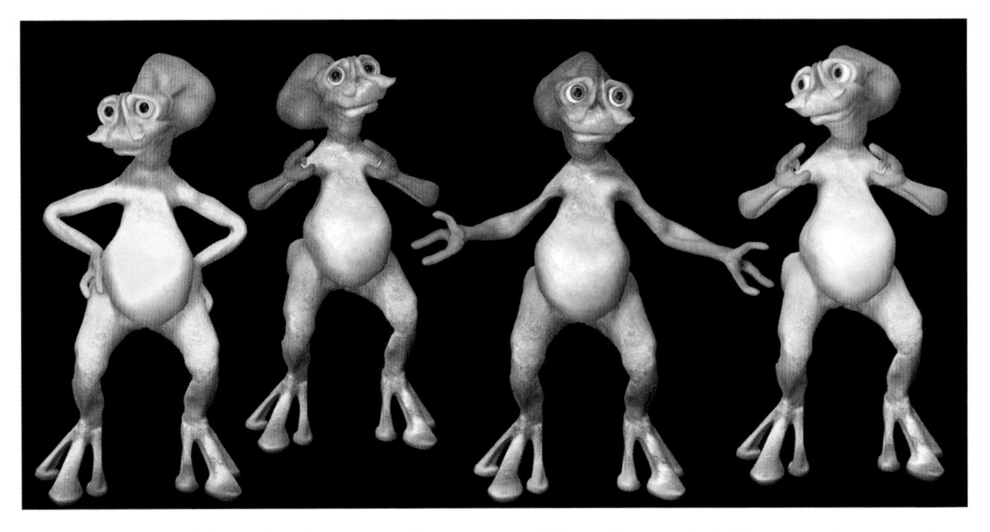

What colour would an alien be? It hardly matters, although most TV aliens opt for neutral grey.

then it will also have approximately the same gravity. Jo will be limited in maximum size by these scaling laws. If his muscles and bones are comparable in strength to ours, he won't be much larger than the biggest land animals on Earth.

Mind you, there are some ways around this limitation. Jo Alien could be truly giant size if he lives in water, a thick medium that makes him relatively "weightless." He might not even require a skeleton, since even large, floppy creatures can survive and maneuver in water. (The giant, boneless squid *Architeuthis* is known to reach lengths of 17 meters.) But would a massive marine alien be smart? Some biologists claim that the ocean isn't a very challenging environment: the weather's the same from day to day, and as a swimmer, you don't have to climb hills or maneuver around rocks. They suggest that this situation might forever fail to encourage great intelligence in sea creatures. The trouble with this argument is that whales, dolphins, and even squids are among the smartest creatures around (although note that the whales and dolphins were land mammals that returned to the sea millions of years ago).

But even if sophisticated sea dwellers exist, it's unlikely that they would develop science and technology. Industrial processes – for example the fabrication of new materials – would be difficult in salt water. Ocean inhabitants might never discover the stars (it's tough to study the sky from underwater) or for that matter, invent radio communication (most radio waves don't penetrate water). So even if there are smart, submersible aliens, it's unlikely that we will ever find them with our SETI experiments.

In summary, it's very unlikely that an intelligent alien would even vaguely resemble us. But if it evolved on a planet similar to ours, there are some aspects of his construction we can reasonably expect – a size somewhere between that of a mouse and an elephant, for example. And we would probably know which bit of the alien to look at and which bit to shake if we ever met. But design aside, is intelligence itself a likely feature of life on other worlds? It's time to address this controversial question.

CHAPTER 3

Intelligent life

In August, 1996, some NASA scientists announced that they had found evidence for life on Mars. This was an astounding claim, and newspapers headlined it in impressively large type.

Mind you, this life was well and truly dead – no more than microscopic fossils from a meteorite that dated back more than 3 billion years. But the implications were important. If two planets in our Solar System have spawned life, then earthly biology is not a miracle. The Universe must be crawling with life.

Nonetheless, within a week of this announcement, the general public lost interest in the story. This happened partly because other researchers disputed the claims; many of them doubted that the squiggly forms seen in the meteorite from Mars were ever alive. But there was another reason that this story didn't set the world on fire: microbes are, to most folks, not particularly captivating. When it comes to alien life, we're more interested in the intelligent variety.

This isn't puzzling. There are certain types of creatures that just naturally get our attention. We have an endless fascination with large predators, for instance. When the so-called educational TV networks want to boost ratings, they churn out yet another documentary about sharks, snakes, or big, hungry cats. For our ancestors, knowing the behavior of dangerous predators was important, as they occasionally had to deal with them face to muzzle. Those humans who didn't have any interest in these beasts were relentlessly removed from the gene pool.

For similar reasons, we're interested in our *intellectual* peers. If we encounter a creature as clever as we are, then there's the disturbing possibility that it will become a competitor for food or territory. On the brighter side, it may turn out to be a potential mate. Fictional aliens routinely show up to either trash our cities or abduct us for breeding experiments, and given our natural survival interests, these behaviors shouldn't be a surprise.

SETI, the Search for Extraterrestrial Intelligence, is a modern-day expression of our interest in high-IQ beings. SETI researchers aren't actually looking for the aliens themselves; they're hunting for evidence of their technology. To do this, they scan the skies for radio or light signals that would tell us that something clever enough to build a transmitter or a high-powered laser was out there. These experiments try to detect intelligence hundreds or thousands of light-years away, and that considerable distance rules out immediate concerns that the SETI quest might be dangerous.

But given that any signals we might discover would be from life we will probably never meet, why bother?

To begin with, we could learn something. After all, we're not going to find signals broadcast by civilizations that are less advanced than we are. Sure, there may be alien societies that resemble Europe during the Roman Empire, or America when the Aztecs were at their peak, but we won't find them. Those societies didn't build radio transmitters. Anyone that we discover will be at our technological level at least, and more likely beyond it (after all, what are the chances that the history of their planet parallels ours to within a few years?). Keep in mind that while the Earth is 4.6 billion years old, the Universe has been around for 14 billion years. There has been plenty of time for beings in other star systems to gain a substantial lead on us (possibly billions of years).

So if some of these advanced societies are kind enough to beam their accumulated knowledge and wisdom our way, we might learn a great deal. Even if they sent us only their culture – their art, their music, or the particulars of their religion (assuming they have such things) – we would find it interesting, and worthy of a traveling museum exhibit. Aliens would suddenly become very trendy. In the last part of the eighteenth century, the native culture of the South Sea islands – recently discovered by Captain James Cook – was all the rage in England. We can expect the same thing to happen for any alien society we might find.

That would be interesting, of course. But bear in mind that if Jo Alien is located hundreds of light-years away or more, two-way conversation is not really practical. We'll only know what he tells us. In addition, and as will be discussed later in this book, it's entirely possible that we won't understand his signals at all. In that case we'll only have the bare fact that we're not alone. That is, in itself, of great philosophical importance. But knowing that there is other intelligence out there would also answer a question that biologists find especially vexing: how likely is it that a planet with life will produce *intelligent* life?

The road to intelligence

There are millions of species on Earth. Millions. Among this protoplasmic plentitude, how many species are smart enough to be interesting on the telephone or able to help you with Sunday's crossword? Well, there's *Homo sapiens*, and then there's . . . nobody.

Is this a momentous fact or not? Is the circumstance that we can look around and find that we're the brainiest boffins on the planet merely a trivial result of being the first species *able* to notice? Or is there some reason to think that intelligence is actually a rare and unlikely evolutionary development, and *Homo sapiens* is just a fortunate accident?

This is more than simply another good question to bandy about after dinner, between the cigars and the port. It goes right to the heart of our place in the Universe. And it's also of obvious and critical importance to SETI researchers. After all, we're on a fool's mission deploying our SETI telescopes if there's no intelligent life out there.

So how can we judge whether intelligence is a likely evolutionary development or not? We do the obvious, and look for hints in Earth's history. Earth is, after all, the only example we have. If high-IQ beings appeared here, and we assume that our planet is just another, typical, run-of-the-mill rocky world, then we could assume that what happened on our planet might happen on theirs, too. Sooner or later, intelligence will arise.

This may be a very naive assumption. The idea that biological history on Earth is simply the flowering of a "tree of life," with humans eventually and inevitably budding at the leafy crown, is popular with the public, but considerably less so with evolutionary biologists. Sixty-five million years ago, a rock the size of London slammed into the Earth, and its catastrophic aftereffects wiped out three-fourths of all species, including the dinosaurs. If this hadn't happened, the rat-like mammals that eventually evolved into *Homo sapiens* wouldn't have inherited the world. That rock could easily have missed us. And 245 million years ago, another catastrophe – the dramatically named "Permian extinction" – wrote *finis* to an even larger

Life can get started and then suffer a natural catastrophe.

percentage of species. If these disasters hadn't taken place (and they were, after all, the result of cosmic chance), then you wouldn't be reading this book. *Homo sapiens* would never have appeared.

So it's certainly possible that in the vast star fields of the Galaxy, worlds with life are abundant, but the inhabitants of those worlds are stupid – the alien equivalents of dinosaurs, worms, or cockroaches.

However, while some biologists think that the evolution of intelligence is unusual and rare, there's another point of view – one that suggests that, no matter how many forks life encounters in the evolutionary road, sooner or later it will arrive at thinking beings. This more upbeat opinion is motivated by a recognition that there are some common animal behaviors that seem to favor the development of intelligence, behaviors that might lead to brainy beasts on many worlds.

Social interaction is one of them. If you're an animal that hangs out with others, then there's clearly an advantage in being smart enough to work out what the intentions of the guy sitting next to you are (before he takes your mate or your meal). And if you're clever enough to outwit the other members of your social circle, you'll probably have enhanced opportunity to breed (to put it graciously), thus passing on your superior intelligence. Another tactic we see among simians involves the sharing of food. If a male is intelligent enough to catch a meal, and canny enough to share it with the females in his group, he will have a better chance of winning their favor and getting more like-minded offspring into the next generation. Higher intelligence translates into higher reproductive success.

Predator–prey relations are another type of interaction that can ratchet up intelligence. When a lioness catches a wildebeest, she's more likely to snag one of the dumb ones that isn't paying attention. Result? The lioness

Advertising your presence can be a good way to get eaten out of the gene pool.

has a meal, but the average IQ of the wildebeests has been raised. This puts the lions at a slight disadvantage in running down their next dinner, and the not-so-smart cats will go hungry and eventually drop out of the gene pool. Both predator and prey will be under selective pressure for intelligence and, over time, both will get brighter. Old World monkeys have far larger brains, relative to their body size, than their cousins in South America. This is because the South American simians had to face only low-grade predators. They were protected from the more aggressive North American hunting animals until recent tectonic events connected North and South America, and provided a door to let in the hungry North Americans.

Of course, these factors alone aren't enough to guarantee improved IQ. Ants are social, and they're also prey to many animals. But they're not smart. The reason is clear: their brains are the size of pinheads. Most researchers agree that, while brain weight may not be everything, a certain minimum number of neurons is necessary for intelligence. In general, bigger really is better.

Indeed, for most of Earth's biological history, big brains weren't in the cards. For billions of years, life was small – mostly single-celled. And even when it progressed to multicellular creatures, intelligence didn't spring up very quickly. As we've noted, a hefty cranium requires a major energy source. Only after oxygen-breathing animals appeared was there a chance for such a development, since oxygen offers the possibility of a high-octane metabolism. But there's more. Animals that spend a lot of time caring for their offspring obviously reap more benefit from intelligence. Learning from your parents is valuable for survival. But if Mum kicks you out of the nest within a few weeks (as occurs for most creatures), then there's a limit to that value. There's little evolutionary incentive to produce intellects that are greater than can be used. So there's no doubt that the appearance of high-IQ creatures required a few pre-existing conditions: an oxygen atmosphere, and a social milieu that permitted extended childhoods.

But while such conditions could produce the alien equivalent of a forest ape, would it inevitably lead to human-style intelligence?

The evolutionary path to *Homo sapiens* gives us some insight into how to answer this question. The heft of hominid brains over the last 3 million years has increased by about a factor of four, but the increase has not been steady. There was a big jump roughly 2 million years ago, when the upright *Homo habilis* made the scene, and then again a few hundred thousand years ago when *Homo sapiens* appeared. The relatively sudden change in brain size suggests that there were evolutionary pressures beyond those we've discussed so far working to make our ancestors smarter.

While we still don't know for sure what these pressures were, one possible scenario starts with a simple change in climate. A few million years ago, the slow drift of continents changed the pattern of ocean currents in the southern hemisphere. East Africa became dry, and our ape-like ancestors found that their familiar, forest habitats were gone. They had little choice but to range across the savannahs in search of food. When they found a meal, it was a tough job to carry it back in their mouths, as primates don't have heavy-duty jaws. Using their two front appendages was a better idea, and evolution favored those who were able to bring back dinner by walking upright. The freed-up appendages, known today as arms, turned out to have other benefits: they were useful for fashioning and using tools. In addition, arms could be used to give "hand signs" (something still frequently and often rudely-done in traffic), and these led to grunts, growls . . . and ultimately English. The emergence of language was probably intimately tied to improved IQ. Try to think about *anything* without forming words in your brain. This will quickly convince you of the link between language and intelligence.

Another possible source for our impressive IQs is the dating behavior of our distant ancestors. You may have noticed that the males of many bird and mammal species deck themselves out with vibrant plumage, appendages, or other coverings to attract the attention and favor of the females. The males display, and the females then decide. Peacocks are an oft-cited example: when choosing a mate, peahens prefer males with long, bright tails. This isn't simply due to a quirk of their little pea brains; it's good reproductive strategy. Those impressive tails are metabolically costly, and require both good health and success in finding food. In addition, a flashy tail can attract predators, which only the wily can avoid. So when a peahen spies a male with a well-endowed tail, she can be sure he has good genes. If she chooses to get to know him better, their offspring will have a survival advantage.

So how does this fit with the evolution of intelligence? It's possible that the ramping up of hominid IQ was the result of a hundred thousand generations of pre-human courtship operating on fitness signals made possible by brain power. A brain is a very complex organ, and can be easily messed up in an individual with slight genetic mutations. The facts are that a good brain is the signal for good genes. A good brain is our flashy, peacock tail.

But it's a messy business, and a real social blunder, to physically display your brain in polite company. So how can a quality cranium make its superiority known to a mate? It does so with behaviors such as speaking well, or by demonstrating musical ability, a sense of humor, or creativity. These activities depend upon many parts of the brain, and consequently are reliable indicators of mental merit.

So males strut their stuff by crooning, being witty, and speaking well, while the females use these clues to sort out the best one to take home to

Did intelligence arise as part of the mating game?

Mom and Dad. You might wonder why this mechanism would ever lead to females who are also clever. But that's a natural consequence of the long maturation time of humans. Since human babies take so long to grow up, there's a lot of evolutionary selection for females who are interesting and amusing, for these will be able to keep their flashy male mates hanging around to help with the kids.

Signaling for fitness, as this mechanism is called, could have been an important process in the rapid appearance of human intelligence. But it's the type of process that might take place in any environment where complex, social animals have appeared. The point is that once the pre-conditions were in place, nothing especially remarkable seems to have been necessary to cause our rapid emergence from just another furry ape to the brainy guy in the next office. It certainly sounds as if Nature – whether on our planet or some alien world – will stumble into increased IQ sooner or later.

Another sort of intelligence

However, even if this is wrong – even if the evolution of biological intelligence doesn't happen often – the Universe could still be chock-a-block with smarts. Let's suppose, at least for the moment, that the pessimists are right: the circumstances that led to intelligence on Earth are, indeed, incredibly unusual. In that case, it seems we would be members of a very exclusive club – the intelligentsia of the Galaxy. We would share the Universe with very few thinking beings.

Or would we? We've talked about Jo Alien on the assumption that he's a hunk of pulpy protoplasm, a mass of living, breathing, tissue. But maybe not. To understand the alternative, consider the future of *Homo sapiens*. In the last several million years, the size of our brains has increased rapidly. But the growth in brain weight has stopped. We haven't gotten any smarter for about ten thousand generations now, and there are good reasons for thinking that this stagnation may continue. For example, modern medicine and enlightened governments allow everyone, and not just the mental heavyweights, to successfully raise offspring, so we've stopped evolutionary selection based on intelligence. Another limitation to increased brain power is the fact that the female pelvis already has a hard time accommodating the birth of babies. Make those infant heads bigger, and you'll cause a real delivery problem.

Beyond that, hefty heads are dangerous. The movie alien *ET* had a thin neck and a head the size of a watermelon. If a real creature were built this way, he would snap off his head the first time he twisted around to check for passing traffic. Much bigger heads on humans would pose a daunting mechanical challenge.

But that may not matter. It's possible that the next step in evolution on Earth won't be improved models of *Homo sapiens*, but something entirely

Will we invent our successors? Has Jo already done so?

different: thinking machines. Technically inclined folks are fond of predicting that sometime in the early part of the present century we will succeed in building a computer that thinks – not simply plays a good game of chess, but really *thinks*. Such a machine could do basic research in physics, perform stand-up comedy, teach your children . . . or edit this book.

Now the interesting point is that this development may already have taken place on other worlds. Even if the emergence of intelligent aliens is an infrequent occurrence, if it's happened at least *once before* in the Galaxy, then those extraterrestrials may have already constructed thinking machines. This would dramatically change the course of evolution on their planet, since cogitating computers can quickly improve themselves. For example, if a thinking machine wanted more memory, it wouldn't have to hang around for a thousand generations in the hope that natural selection would improve its brain. After all, with Darwinian evolution, you can't guarantee what's coming next. More memory might happen, and then again,

There are limits to what biological creatures can do, and thinking machines could quickly become superior.

it might not. But as a machine, all you would need to do is add a few more chips to your motherboard. In the world of machines, individuals can evolve, not just species.

Clearly, once thinking machines are built, they can quickly overwhelm the abilities of biological intelligence. It's been estimated that the raw computational power of the human brain is somewhere between ten trillion and ten thousand trillion operations per second. The dumb computer on your desk is at least 10,000 times *less* capable. But computers are doubling in speed every 18 months, which means that in another two decades, an office computer will be more agile than the person seated in front of it.

So imagine the capabilities of thinking machines originally built by a galactic civilization that arrived at our technological level a billion years ago or more. Such advanced intelligence might not be happy just to stay in the solar system of its birth. And unlike biological beings, machines (not being mortal) wouldn't balk at the idea of spending thousands of years rocketing between the stars. They could do that. And presumably, some of them have.

This suggests that the bulk of the intelligence in the Universe might not be soft and squishy at all, but made up of highly refined, extraordinarily advanced thinking computers. What would be the interests of such machines? Would they be content to live in the company of their (relatively primitive) biological creators, or would they plunge into the vast voids of the Galaxy in search of new knowledge, or simply new sources of material and energy? Would they need company, a "society" of other machines in which to function? Or would they simply be immense, solitary masses of brainy hardware, happy to drift through the darkness of interstellar space, and occasionally signal their presence or their thoughts to other machines?

It may be that as wonderful as biology is, it's only a primitive bootstrap to producing the true masters of the Universe.

CHAPTER 4

Visitors from afar

Is it possible that questions about where the aliens come from, what they look like, and whether they're animals or automatons could be answered by stepping into the backyard? Is it possible that the aliens are already here on Earth?

A lot of people think so. Public opinion polls in the United States routinely show that more than half the citizenry believes that the American government is hiding information about visiting extraterrestrials. The popular opinion in Europe is similar. According to many people, the aliens are buzzing our airspace and occasionally molesting the citizenry.

Many believe that the aliens are already here.

Clearly, if this is true, then there are at least some people who really know what Jo Alien is like. They have the answers to questions that humans have asked for thousands of years.

Is this reasonable?

Why people think they're here

More than two thousand years ago, the Greeks were telling stories of gods that came down from the heavens to mess with the human populace. The idea of visitors from space is an old one, for, after all, the sky is dark, mysterious, and undoubtedly a great place to hide all sorts of powerful creatures.

The modern version of the Greek tales began in 1947, when an Idaho businessman named Kenneth Arnold told newspaper reporters that he had seen a clutch of strange objects while flying his small plane. They were, Arnold claimed, bowling along past mountain ranges in the American northwest at 2,000 km/hr. They darted hither and yon, moving "like saucers skipping in water," a description that was misunderstood by the reporters. They wrote up the story stating that Arnold had seen "flying saucers" (in fact, he thought the objects were crescent-shaped, but "flying crescents" sounds more like a trapeze team).

Suddenly, flying saucers were being seen everywhere. A captain in the US Air Force soon coined the term "Unidentified Flying Object", or UFO, a name that is frequently used today for any bright light or object seen in the sky that's not easily identified.

The assumption made by many people (often before looking at the alternatives!) is that at least some of these UFOs are interstellar spacecraft, with aliens in the cockpit. Of course the question naturally arises, *why* would aliens visit us? In science fiction films, the answer to this is easy. Extraterrestrial visitors usually show up on the doorstep to trash our planetary home, or at least take it over. However, as the astute reader will notice, Earth hasn't yet been destroyed, despite a half-century of UFO sightings. So other motives for alien visitation have been put forth.

Perhaps these other-worldly beings want our resources. But which ones? It has been occasionally suggested that the asteroid belt, with its relatively small chunks of concentrated metallic ore, might be an appealing export item for aliens, encouraging them to visit our neighborhood. Another possibility is that the extraterrestrials want our water or perhaps some of our gold, silver, or platinum. In fact, all such ideas are a bit of a stretch, as it would be considerably less expensive for the aliens to find these materials in their own solar system and save the shipping expense.

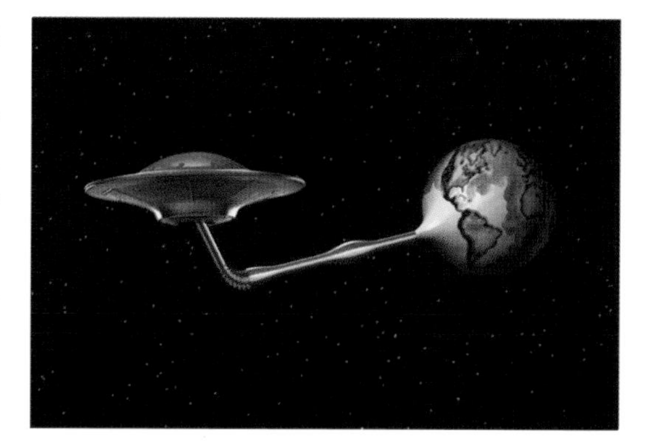

Would aliens want our resources?

Another possibility is that these star people have come to breed with us. This is not only unappealing, it doesn't make much sense. To do this, the aliens would need to have DNA as their genetic blueprint, and beyond that, DNA that was nearly identical to our own. Humans can't successfully breed with chimps (and are usually discouraged from trying), despite the fact that our DNA differs from theirs by less than two percent. In addition, it's hard to imagine that Nature would produce a species that could only reproduce by rocketing off to a distant star system to find other-world partners. That's a difficult dating situation.

. . . our affections . . . ?

If it's not our planet or our bodies they want, what about our technology? This, too, is less than plausible. Clearly, if they have the ability to traverse hundreds of light-years or more to come to Earth, then their technology is far beyond ours. Would we visit cavemen to steal their technology?

A few people have suggested that the opposite has happened: that governments have "reverse engineered" alien spacecraft and thereby made great technological advances – particularly for such items as

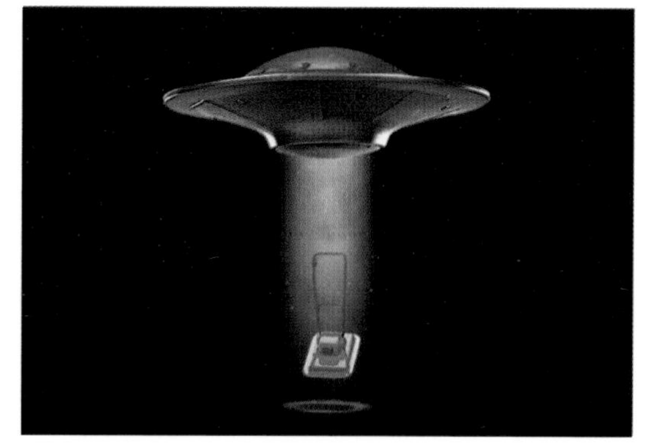

. . . or our technology . . . ?

military aircraft. This would explain the popular idea that the US government has hard physical evidence of alien visitors, but is keeping it secret. Again, this doesn't make much sense. It's unlikely that if you handed a cellular phone to a Neanderthal he would be able to "reverse engineer" it. He'd probably just use it to pound bones. In addition, if this were really happening, you would expect that American military aircraft would be *enormously* more capable than anyone else's, something that's not true.

These explanations for alien visitation are intriguing, but don't seem to be very compelling. Then again, maybe we needn't worry about motives if there's really impressive evidence that extraterrestrials are on Earth. If they're here, that's important and interesting, no matter what their reasons for coming. But how good *is* the evidence?

Sightings and other alleged alien manifestations

UFO organizations all over the world log thousands of sightings annually. Many of these are undoubtedly airplanes, parachutes, birds, kites, meteors, ball lightning, bright stars or planets, rockets going up, rockets falling down, weather balloons, party balloons ... the list goes on and on. But no one would be terribly interested in learning that someone had sighted the 8pm flight from Gatwick to Paris. What gets the blood pounding is the thought that some of these UFOs are alien craft.

Alas, there's no reason to think that's true. Despite considerable efforts to investigate these UFO sightings, particularly by the US Air Force (USAF), there's been no evidence sufficient to convince most scientists that these lights in the sky have anything to do with alien rocket jockeys. The USAF began its study of the phenomenon in 1947, just as Kenneth Arnold introduced the world to the idea of flying saucers. It's safe to say that the military was more concerned that these mysterious objects might be advanced aircraft from the Soviet Union, rather than spacecraft from the stars. Even Prime Minister Winston Churchill was advised of the possibility. But whatever the motivation, the USAF checked out nearly 13 thousand UFO sightings between 1947 and 1969. Of these, the overwhelming majority – 94 percent – were eventually identified as being due to natural or human causes (planets, aircraft, etc.) When the Air Force finally shut down its investigation at the end of 1969, it concluded that UFOs were not a threat, nor could it find any evidence that they were extraterrestrial vehicles.

Of course, you might wonder about the 6 percent of sightings that the Air Force didn't identify. Maybe *those* were extraterrestrial visitors? It's an appealing thought, but it's hardly credible proof. The police department in a big city may solve 94 percent of the murders (if the force is exceptionally competent), all of which will involve human perpetrators. But what about the other 6 percent? Could they be caused by ghosts or leprechauns or aliens? Of course they could be, but who would believe that? In fact, it's fair to assume that if the police had more detectives, evidence of better quality, and more time, a greater percentage of the crimes would be solved.

This is the light pattern made by an out-of-focus camera lens. Such optical effects are undoubtedly responsible for many claimed photos of UFOs.

UFO research is similar: there's seldom sufficient time or manpower to explore all the ordinary phenomena, such as astronomical objects or aircraft, that could account for a sighting. While thousands of UFOs have been spotted, they're usually seen either by lots of people from far away, or by small numbers of people (often only one) somewhat closer. This means that the descriptions – and the occasional photographs and videos that accompany them – are either rather poor or are hard to confirm. Many reports just don't have enough information to make an identification possible – no matter what the object might be – so it shouldn't surprise you that not all UFOs have been converted into IFOs (Identified Flying Objects).

In addition, there are a host of Earth-surveillance satellites that constantly watch our planet from above. Some of these are for military purposes, while others are used to monitor the weather or to map vegetation, ocean currents, or even volcanic eruptions and fires. They don't find any alien craft. And in the United States, the military has operated a sophisticated radar network to find and catalog space junk for more than four decades. It keeps tabs on thousands of pieces of debris, some as small as a cricket ball. This radar doesn't seem to find alien craft either. Somehow,

An alien intent on abduction bags its prey.

the visitors can be seen by ordinary folk standing in their backyards, but not our high technology.

Aliens up close

Most UFOs are observed while they flit across the sky. This naturally prompts one to ask: why don't they ever land? When the Europeans first explored the Americas, they didn't simply sail a few miles off the coast, never bothering to make it to shore and engage the natives face to face.

However, there is at least one phenomenon suggesting that the aliens *do* land. These are the reports by people who claim they have been abducted from their homes. (They are usually removed from their bedrooms; the aliens seem reluctant to take folks from the dinner table or the TV couch.) The abductee is then spirited aboard spacecraft for observation or uncomfortable experiments. A poll taken in the United States in 1992 suggested that more than three million citizens think they have had a UFO abduction experience. If this is true, it indicates that the aliens are quite busy. However, researchers have pointed out that such experiences are an old story. Even in the Middle Ages, reports of abductions during sleep occurred, although in those days it was usually the Devil or one of his representatives who was thought responsible. Psychologists have pointed to a phenomenon known as sleep paralysis that might naturally account for these stories. When we're asleep, and even for a short time after waking, our brain arranges things so that we have difficulty moving our bodies (which is a good thing if you occupy the top level of a bunk bed). In this circumstance, we occasionally experience a dream state in which we feel unable to respond – in which we're frozen in bed and have visions of being watched. This phenomenon has been offered as a possible explanation for some of the reported abductions.

In addition, there is at least one widely publicized case in which the aliens are said to have landed – rather disastrously. This is the so-called Roswell incident. In mid-June of 1947, the same month in which Kenneth Arnold made his saucer sighting, a rancher about 100 km northwest of Roswell, New Mexico, in the American southwest, found some unusual debris on his property. He reported this to the military, which drove out to the ranch and promptly removed the debris, explaining it away as a "flying disk." Within a day, the military-powers-that-be changed their story, saying that the collected material was merely a weather balloon.

It didn't look much like a weather balloon, and it wasn't. Years later, the US Air Force changed the story again. In a lengthy 1994 report, it noted that the Army Air Field at Roswell had been busy in the late 1940s with a highly classified experiment known as Project Mogul. Mogul was designed to monitor nuclear tests in the Soviet Union by launching constant-altitude balloon trains near its borders. These would have sensitive listening devices able to pick up sound waves produced by a distant blast. As part of the testing for Project Mogul, balloon-borne equipment was being launched from the air base near Roswell, including a June 4 lift off consisting of nearly two dozen balloons tethered together in a long line. Later analysis showed that this particular launch would have drifted towards the ranch where the debris was found. According to a surviving member of the Project Mogul team, both the appearance and content of the debris matched well with what the researchers at the air base were lofting into the air.

In other words, there's a reasonable, if somewhat complicated explanation for what was found at Roswell that involves a US military experiment, but not aliens.

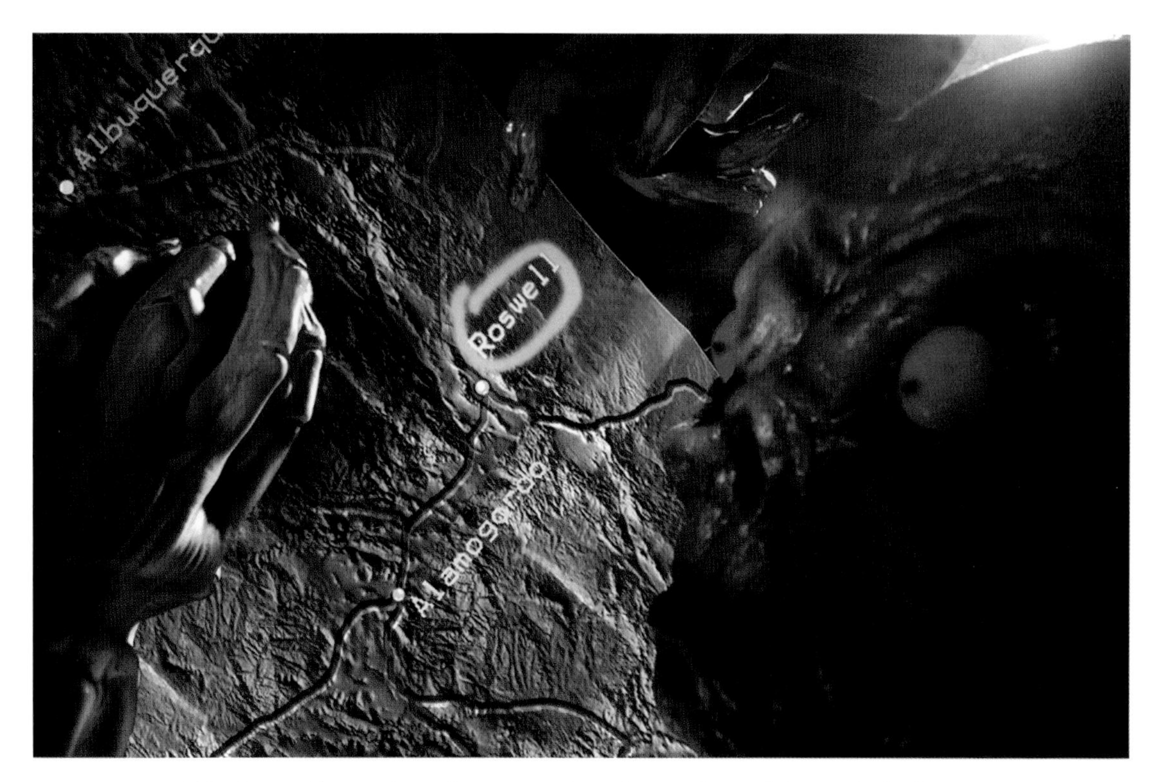

Did aliens suffer a mishap in 1947, and crash near Roswell, New Mexico?

There are other supposed signs of alien activity on Earth. These include the Egyptian pyramids and the unusual, and ancient, markings raked in the desert near Nazca, Peru. But it's difficult to believe that the giant pyramids of the pharaohs were built by alien visitors, simply because there was an obvious progression in the sophistication of these structures as time went on. For example, the famous step pyramid at Sakhara, built five thousand years ago, partially collapsed because it was too steep for the weight of the blocks used in its construction. The famous Giza pyramids, younger by a millennium, had smooth (non-stepped) sides and more gently inclined slopes. They were a substantial refinement on the construction at Sakhara. This sort of evidence strongly suggests that the Egyptians were learning how to build these monuments, rather than that they had hosted alien architects sophisticated enough to travel between the stars, but too clumsy to stack up blocks at a supportable angle.

The so-called Nazca lines consist of kilometers-long, straight and nearly parallel lines scratched on the Peruvian desert, as well as oversize geometric forms and animal figures (a turkey, spider, hummingbird, monkey, etc.) It has been suggested that those were constructed by local inhabitants following alien instructions. The aliens were apparently eager to have landing strips for their saucers. But as with the pyramids, it's hard to understand why alien intervention would be necessary. Archaeologists, who believe that the Nazca markings date from about the time when the Roman Empire was flourishing in Europe, have put forth a few reasons why they might have been built: either as components of a religious ritual, or to mark sources of water. But whatever their function, the local Peruvian inhabitants of that time were certainly sophisticated enough to push rakes across the desert to create the markings, and it's a bit condescending to think that they needed extraterrestrial help.

Finally, there are the enigmatic and decorative "crop circles" that appear every summer, most frequently in England. Despite the fact that these can be produced by motivated teams of artists using nothing more sophisticated than boards and ropes, there are still folks who are convinced that some of these grain graphics are produced by more exotic draftsmen – visiting aliens, intent on communicating to us with signs. Of course, one has to wonder why the aliens would direct so much of their communication to a few rural counties in England, and even more, why any advanced society would be willing to spend the enormous amounts of energy required to traverse the distances between the stars merely to carve graffiti in our wheat. In addition, while attractive, the crop circles usually have repetitive, highly geometric designs. In the language of signal theory, they convey very little information (far less than the words of this paragraph). It seems bizarre in the extreme to think that sophisticated extraterrestrials would go to so much trouble to tell us so little.

Is there a reason for the lack of good evidence?

Rather few scientists are convinced that aliens are visiting Earth. The foremost reason they give to justify their skepticism is the lack of hard physical evidence. There are no obvious artifacts. No one ever shows up at a university lab with a piece of unknown plastic or a weird alloy from a UFO – something that could be analyzed.

Crop circles are real. The argument is over what or who causes them.

But those inclined to believe that extraterrestrials really are here have two handy arguments to rebut the skeptical scientists: (1) the evidence is actually more convincing than the scientists realize, but they just are too close-minded to look at it, and (2) the best evidence has been covered up by the (American) government.

The first explanation is hard to swallow. There are hundreds of thousands of researchers in the world. If they thought there was even a tiny chance that aliens were roving this planet, thousands of them would be working on this problem in their attics during their spare time. After all, what could be a more exciting discovery?

As for government cover-ups, this too is a flimsy explanation. Wouldn't every government in the world have to be part of the cover-up? Just because

Maybe aliens are already observing us.

the Americans or the British decide to keep the aliens secret, would the Belgians, Brazilians, and Bulgarians all go along? Or do the aliens only visit countries with secretive governments?

Finally, we might ask, if extraterrestrials really are cruising our skies, why have they chosen to do so now? As we've noted, the first UFO sighting was in 1947. That's five decades ago in a planetary history of 4,600,000,000 years. If we assume for the moment that these claims are real, this chronology tells us immediately that (1) we are the beneficiaries of an enormously rare event (one chance in 100 million), or (2) the aliens routinely visit Earth, or (3) our activities (nuclear tests, environmental degradation, rock music, etc.) have attracted the aliens' attention, and encouraged them to drop by.

The first possibility, that we just happened to be fortunate (being around for the first and only alien encounter), is less probable than that you – not someone, but *you* – will win next month's lottery jackpot. It strains credulity, to use polite vernacular.

The second possibility, that Earth hosts extraterrestrials on a routine basis, and therefore a visit during your lifetime is not particularly improbable, deserves a bit more scrutiny. The question is, how often do they visit? If it's only once in a few tens of millions of years, we're back to the first possibility, and the odds are highly stacked against you being one of the lucky visitees. But as we've noted, some folk claim that aliens have glided to Earth in historical times (to help build the pyramids, for instance). If any of this is true, it argues for visits at least once every 5,000 years or so. The problem with this line of reasoning is that – barring some reason for them to visit humans in particular (a possibility we consider below) – it implies that there have been at least a *million* expeditions to Earth! We may send the occasional anthropological research team to Borneo, but we don't send a million. And it's a lot easier to get to Borneo than to traverse hundreds or thousands of light-years. This, too, seems to be an unlikely explanation for visitors now.

Finally, we consider the last possibility – we have enticed the aliens with human activity. Let's set aside the question of whether advanced galactic societies would have the slightest interest in our wars, our pollution problems, or our popular music. The real question is, how would they know about us at all?

In fact, there's only one clear and persistent "signal" that *Homo sapiens* has ever sent to the stars: our high-frequency radio transmissions, including television and radar. The Victorians (let alone the Egyptians or the Nazca Indians), despite all their technical sophistication, could never have been spotted from light-years away. Humans have only been making their presence known to the Universe for the last 70 years.

And that's a problem. It means that even if the aliens can immediately scramble their spacecraft after receiving one of our first broadcasts, and fly to Earth at the speed of light, they can't be farther than roughly 8 light-years away to have arrived by 1947. (The first high-frequency radio transmissions from Earth – the type that would leak into space – began in the early 1930s.) There are four star systems within 8 light-years distance. Count 'em, four. We're back to winning the lottery.

What about warp drive? Maybe the aliens can create wormholes and get here in essentially no time. It doesn't matter. Our signals travel at the speed of light, and this means that even with infinitely fast spacecraft, the aliens can't be farther off than about 15 light-years to have reached our lovely planet by 1947. The number of star systems within 15 light-years is about three dozen. There would have to be 10 billion technically sophisticated societies in the Galaxy to have a reasonable chance of finding one camped out among the nearest three dozen stars. That's optimism of a high level indeed.

It's nice to think that either Earth or its human inhabitants have not only attracted the attention of galactic neighbors, but encouraged them to visit. But frankly, the numbers don't give much support to this somewhat self-centered idea.

We want to believe

The sheer number of UFO sightings, crop circle appearances, and abduction reports gets our attention. But big numbers alone aren't enough to prove alien visitations. They may, however, tell us something about human behavior.

We like the idea of powerful beings, able to influence our personal lives. In a Universe that's incredibly vast, bitterly cold, and brutally hostile – a Universe in which Earth is an unnoticed dust mote drifting nowhere in particular – it's comforting to think that there are beings that have managed to master the forces of Nature sufficiently to find our planet, detect us, and – lo and behold – come here to affect our mundane existence. Although they may seem mostly keen to probe our bodies and use us for breeding experiments, at least they're showing interest. We may be fearful of abduction, but on the other hand, it's flattering that they want us at all.

The other angle to this is the fact that the UFO phenomenon is "people power." Mainstream scientists do not believe that there are aliens wandering the landscape. But for most folk – unfamiliar with the subtle concepts and difficult mathematics of modern science – the claim that aliens are visiting is empowering. This is one area, a deeply important area, in which they can claim that they know more than the academics. They have *seen*

This picture was made in response to a request by a magazine for a "real" UFO photograph. The saucer is merely a lamp shade.

the aliens; they know the answer to an important question that the experts have not yet answered. It's David *vs* Goliath, and Goliath is the arrogant science establishment (or so it is perceived). This seems to be why so many members of the general public get positively exercised over the question of whether the Roswell incident really involved aliens or not, or if lights in the sky are craft stuffed with small gray guys.

But in the end, the evidence will provide the answer to these questions. And so far, the evidence that they're here isn't very impressive.

CHAPTER 5

How might we get in touch?

We've discussed both the likelihood and the possible appearance of extraterrestrials. But all our speculations will remain no more than careful guesswork unless we can find evidence that the aliens are out there.

Unfortunately, finding the evidence might be hard. Obvious clues to cosmic company won't exist unless the aliens manage to travel through, or signal across, the distances between stars. And those distances are substantial. Astronomer Friedrich Bessel became the first person to actually get a handle on interstellar expanses when, in 1838, he determined how far it was to the faint star 61 Cygni. He measured a distance of 10 light-years,[1] or 100 trillion kilometers. This is an incredibly long way: 200 million times farther than the Moon. And 61 Cygni is one of the *closest* stars. The Universe is depressingly large, and consequently even a reconnaissance of our cosmic backyard has been – until recently – nothing more than a pipe dream.

But it may no longer be. Technology developed in the last five decades has finally given us the possibility of contact with extraterrestrials, despite the distance. Today, a small group of researchers is trying to turn that possibility into fact.

Let's just go there

As we discussed in the previous chapter, it seems doubtful that any aliens have actually come to Earth, at least in modern times. But if the aliens haven't visited us, is it possible that we could visit them? This idea is perennially popular in science fiction. From *Star Trek* to *Star Wars*, humans boldly go where they haven't gone before – in search of extraterrestrials.

[1] In fact, the modern value is somewhat larger, about 11.2 light-years.

The Moon, more than a hundred times closer than Mars, is the only cosmic body yet visited by humans.

This "greetings card," mounted on the Voyager spacecraft, also contains instructions on how to play the LP record inside. How many people have access to a record player today?

The trouble with this idea is that going boldly isn't the issue. Going *at all* is the issue. The farthest into space that humans have ventured is the Moon – a mere hundred times the span between London and New York. You'll probably commute a greater distance during your working lifetime. And yet, sending white-suited astronauts to the Moon required a massive effort and an impressive pile of cash. We've propelled robot spacecraft to Mars, but launching *humans* to the Red Planet is such a tricky and costly endeavor that it may not occur for another two decades.

Our fastest rockets cruise at about 15 km/s. At that speed, you could reach Pluto in about a dozen years. But a trip to 61 Cygni would consume more than two thousand centuries – which is longer than you'll want to sit

Rocketing between the stars is a nice idea, but not as easy as movies suggest.

in a cramped seat eating microwaved dinners. This is something to keep in mind when you consider how effective it is to try and contact the aliens by attaching plaques or records to our spacecraft. The Pioneer 10 and 11 craft, as well as the two Voyager probes, launched in the 1970s to study the gas giant planets of the outer Solar System, all carried such greeting cards. The messages were simple and, one hopes, informative, consisting of pictures of us and our planet, as well as a tasteful selection of music and voices. But the time required for these snail mail communications to reach the stars is enormous, and the chances that they'll ever be fished out of the black ocean of space is small.

Going to the stars with our present hardware requires industrial-grade patience. The obvious way out of this discouraging travel situation is to make faster rockets. Of course, that's possible. But *really* fast rockets – the kind that might bring us to nearby stars in less than a human lifetime – are mightily difficult to put together. Many among the public, raised on the flawed fodder of television sci-fi, assume this is just an engineering problem; all that's needed is to wait for better technology. Eventually we'll be able to build spacecraft of any speed. Unfortunately, physics gets in the way of this sunny scenario. To reach the nearest stars within 50 years

A futuristic spacecraft propelled by laser power.

requires rockets capable of coasting at 10 percent the speed of light or more. The energy needed to kick a 100-ton craft up to that impressive velocity (and slow it to a stop at its destination) is as much as the city of London will burn during the next several thousand years, assuming it maintains its current energy usage. More exotic propulsion schemes, such as constructing wormholes in space, are often portrayed in science fiction. But even if such high-speed travel modes are possible – and we're not sure they are – wormhole construction would require unimaginably vast supplies of energy.

A more conservative suggestion is to forget about warp speed rockets, and simply take a leisurely cruise to the stars. Imagine building a very large interstellar spacecraft, filling it with motivated and highly trained folk, and waiting for their distant descendants to eventually touch down on someone else's world. This seems like a foolproof scheme, but many sociologists are unsure whether the crew – even a large crew – would survive after being confined to a space ark for many generations. Experience on

A generation starship taking humans to the stars.

Earth suggests that the people aboard would soon break up into rival gangs. Rather than peacefully cooperating in a noble venture, they might soon be at one another's throats, scheming for control of the craft – and probably resorting to violence. It's also unclear whether the younger generations – who might only read about the ship's mission in archives – would feel motivated to explore new worlds, or would have the skills to do so. In any case, even if interstellar migrations are possible, they don't offer a quick way to find out whether we're alone in the Cosmos.

It boils down to this: we'll not be sending ourselves to the stars in your lifetime. And as we've discussed in the last chapter, there's no compelling evidence that the aliens are in our neighborhood now. So is there any other way to learn whether intelligent beings populate the Cosmos?

There might be. We could look for evidence – either some sort of giant object (the product of a super-size alien engineering project), or signals that distant societies may have either deliberately or inadvertently broadcast into space.

Let's first consider what sort of objects we might find that would betray alien existence.

Alien artifacts

Homo sapiens has been shuffling the surface of the Earth for a few hundred thousand years, and – despite government anti-litter campaigns – has left plenty of debris behind to prove it. Think of the large-scale building enterprises that are spread over the landscape: the ruins of Rome, the Great Wall of China, suburban housing estates. Most of these constructions will disintegrate within a few tens of thousands of years (much sooner for the housing estates), but there are plenty of human artifacts that will last

Finding an object of clearly artificial nature would be strong proof of alien visits.

longer. Aluminum tab tops from soda cans come to mind. Ditto for spacecraft abandoned on the Moon, which – being free of weather – doesn't quickly erode objects on its surface. Could it be that comparable, enduring litter is afloat in the Universe awaiting our discovery? Could we find alien trash?

Clearly, this is a possibility, and since our Solar System is billions of years old, alien tourists from long ago might have left subtle calling cards. Perhaps they've hidden a monolith on the Moon, as Arthur C. Clarke hypothesized in his novel *2001*. Or maybe they've constructed an attention-getting feature on Mars, such as the infamous "Face." We haven't looked very hard

The Face on Mars

Even if aliens aren't gliding through our skies, is it possible that a few could have recently visited the neighborhood and left us a calling card? Some people claim this has happened.

In July, 1976, NASA's Viking Orbiter 1 spacecraft was circling Mars, hunting for a good touchdown spot for the second of the Viking Landers. The preferred location was in the northern area of Mars known as Cydonia. Photos returned from the Orbiter during this reconnaissance showed a terrain dotted with isolated mesas, or table-shaped hills. One of these looked surprisingly like a face.

The NASA scientists figured that the funny looking feature would amuse and interest the public, and a week after the photo was taken, they released it to the press. The text that accompanied the photo noted that "the picture shows eroded mesa-like landforms. The huge rock formation in the center, which resembles a human head, is formed by shadows giving the illusion of eyes, nose and mouth."

To the space agency's surprise, not everyone thought that the face was an illusion. Some maintained that it really *was* a face: an artificial construction intended to get the attention of humans once we were advanced enough to fly rockets to Mars and find it. For the next 22 years, the Face was the subject of vigorous debate on radio talk shows, at sci-fi conventions, and on the Internet. Was it real, or simply an optical illusion? The feud over the Face became more heated than Saharan sand.

Then in 1998, the Mars Global Surveyor took a new picture of the Face, with ten times the detail of the old Viking photo. The picture showed what looked like (to most people) a natural butte. But the argument wasn't over: the new photo was made through wispy Martian clouds, and who knows what might have been hidden? Friends of the Face maintained that NASA was attempting a (literal) cover-up. (Many NASA researchers on the other hand probably wished that the Face really *was* an alien artifact, as that might provide a shot in the arm for the agency's budget.) Indeed, when other images of Mars were examined, some

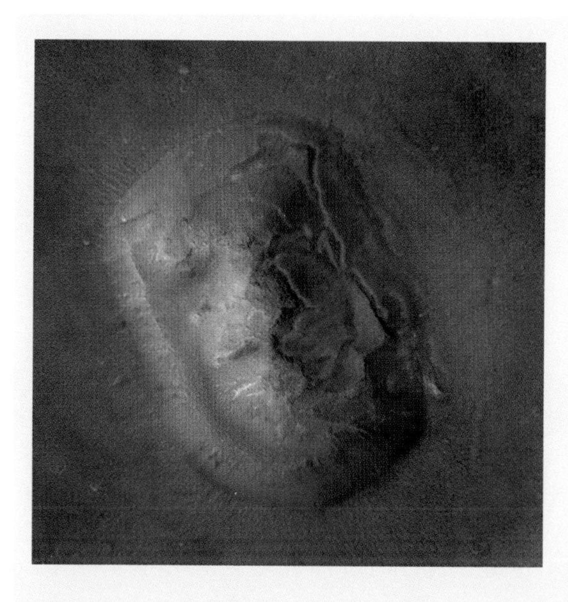

people decided they could see whole cities, pyramids and other structures on the surface.

The idea of a face as a signaling device is strange. If alien architects really built the Face, they would have had to do so in the last few hundred thousand years. *Homo sapiens* is no older than that. If extraterrestrials had visited Mars 100 million years ago, for instance, they would have built a dinosaur face. Isn't it rather remarkable that they just happened to come to Mars once our species appeared?

As it turns out, arguments over the Red Planet's facial feature have recently died down. In April, 2001, the Global Surveyor took a really clear snap of the Face. Detail down to only a few meters in size is visible in the new photo (which is 30 times better than the 1976 image). And a glance at the latest image should convince you that the Face on Mars should really be called the Butte on Mars.

But it's really a beaut of a butte.

for the former, and the latter is almost certainly a natural geological feature, not a piece of alien sculpture. Still, looking for artifacts is a legitimate idea, and various researchers continue to consider whether it might be possible to find alien robot probes in our Solar System, high-powered interstellar rockets in the depths of space, or even massive construction projects in the neighborhoods of other stars.

The problem with the first idea is that the Solar System is a big place, and it's hard to comb it for probes or other artifacts that might be no bigger than a car. Finding interstellar rockets is hard because (1) we don't know where to look, and (2) even the most powerful spacecraft would be difficult to detect. As for alien astroengineering projects – well, what sorts of things would we search for? Clearly, if some advanced extraterrestrials have rearranged the stars in their neighborhood into a precise geometric pattern, that might catch our attention. But then again, it's probably fair to assume that much of what the aliens build would simply be unrecognizable to us.

While a few hunts for artifacts have been made, the search is daunting and uncertain, and most researchers are unclear about how to start. However, this is not the case with another method of discovering the aliens: looking for signals.

Searching for extraterrestial messages

As we described in the last chapter, the most obvious clue that humans exist on Earth – at least from the point of view of distant aliens – is the torrent of radio and television signals that we send into space. Such broadcasts are typically aimed towards the horizon, to reach the spread-out population of a large urban area. But since the signals travel in a straight line, they eventually sail over the roofs (and antennas) of those in the distant suburbs, and head for the stars. Because we've been broadcasting for about 70 years, the earliest transmissions are 70 light-years away, expanding in a large radio-filled sphere that by now has washed over ten thousand star systems.

How far will these transmissions go? In fact, they will continue on indefinitely. But they become weaker with distance, simply because they are filling more space. Indeed, by the time that a typical television signal reaches the nearest stars, it's so feeble that you would need a few thousand acres of unsightly antennas simply to pick it up.

But the point is not whether we're broadcasting our TV culture into space. The point is that *other* galactic civilizations – ones that have been around longer than we have – might be broadcasting theirs. Indeed, some societies may have been transmitting for many thousands of years. In addition, if they're deliberately trying to make contact with other worlds, they would surely use large antennas as mirrors to focus their transmissions – a technique used by radar installations on Earth. By concentrating the radio energy with a big reflector, they can greatly increase the strength of the signal at the receiving end.

In 1959, a pair of physicists at Cornell University, in the United States, worked out the numbers associated with this idea. To their surprise, they reckoned that our most powerful radar sets could send quite detectable signals across a distance of many light-years. To pick up such signals, you could

The 140-ft radio telescope at Green Bank, West Virginia.

use an instrument known as a radio telescope. This is nothing more than a super-size antenna connected to a specialized receiver. On our planet, radio telescopes are used day-in and day-out by astronomers to map the natural cosmic static from such objects as pulsars, quasars, and the gas between the stars. The two Cornell physicists realized that these antennas could also be used to look for alien "radar sets" broadcasting from nearby star systems. They could tell us if anyone is out there.

It wasn't long before someone tried out their idea. In the spring of 1960, a young astronomer named Frank Drake became the first to use a radio telescope to search for extraterrestrials. Drake was a fresh hire at a radio observatory in Green Bank, West Virginia, where a 26-meter antenna had just been installed. The observatory director encouraged his staff to think of experiments that could make use of the new dish. So Drake suggested that one suitable project would be to swing the antenna in the direction of nearby, Sun-like stars, and look for artificial signals – signals that might be coming from planets with sophisticated beings. The director gave him the nod, and Drake spent a couple of weeks listening for alien broadcasts. He got up early each morning to tune the telescope's receiver in the cold, West Virginia air, after which he would point the antenna at the Sun-like stars Tau Ceti and Epsilon Eridani (both of which are roughly a dozen light-years away). In a whimsical salute to writer Frank Baum, he named his short search, Project Ozma, after the princess in *The Wizard of Oz*. Ever since, most SETI experiments have had names that are either prefixed with the word "Project" or are acronyms (e.g. Project Phoenix, Project Argus, Project Cyclops, META, BETA, SIGNAL, and BAMBI).

Drake didn't hear any extraterrestrial signals, but his experiment was the first modern search for far-off aliens. It could have worked. And like most pioneering efforts, it was soon being imitated by other astronomers who had a bit of extra telescope time and a wish to be the one to find ET.

Today's radio SETI searches use instruments that dwarf Drake's first effort, but they're trying to do much the same thing (see Table 1). Many members of the public assume that a bevy of telescopes and a small army of astronomers are cranking away night and day pursuing extraterrestrial signals. The truth is slightly less imposing. The total number of scientists actually scanning the skies is small – a few dozen, worldwide. In addition, most radio SETI experiments run on borrowed hardware: they use

Table 5.1. *Current SETI Experiments*

Experiment	Institution	Telescope	No. of frequency channels	Width of channels	Website
Project Phoenix	SETI Institute	Arecibo 305-m radio telescope	58 million	1 Hz	www.seti.org
SERENDIP IV	University of California, Berkeley	Arecibo 305-m radio telescope	168 million	0.6 Hz	seti.ssl.berkeley.edu/serendip
Southern SERENDIP	University of Western Sydney, Macarthur	Parkes 64-m radio telescope	58 million	0.6 Hz	seti.uws.edu.au/main/serendip.htm
SETI@home	University of California, Berkeley	Arecibo 305-m radio telescope	33 million, using narrowest bandwidth	0.07 Hz and higher	setiathome.ssl.berkeley.edu
Optical SETI at Harvard and Princeton	Harvard University, Princeton University	Oak Ridge 1.5-m and Princeton 0.9-m telescopes	Visible light	Broadband optical pulses	seti.harvard.edu/oseti (Simultaneous observations at Harvard and Princeton)
Optical SETI at Berkeley	University of California, Berkeley	Leuschner 0.76-m telescope	Visible light	Broadband optical pulses	seti.ssl.berkeley.edu/opticalseti

Amateur SETI searches

SETI scientists typically use monster-size radio antennas and research-caliber mirror telescopes for their searches. They do this not to impress their colleagues, but simply because these professional instruments are big – and therefore able to pick up fainter signals. In addition, they are outfitted with exquisitely refined, sensitive detectors.

Given such powerhouse efforts, does it make any sense for amateurs to compete in the search for extraterrestrial signals?

The SETI League has organized radio amateurs to search for extraterrestrial signals using backyard antennas.

In fact, it does. It's no secret that amateurs have made many important discoveries in astronomy. For example, enthusiasts using binoculars and backyard telescopes find the majority of new comets. How is it that amateurs can ferret out so many comets? The reasons are twofold: (1) a lot of them are searching, so they can collectively scan large areas of the sky, and (2) comets can be bright enough that you don't need a big telescope to see them.

The "Very Small Array" – an amateur SETI telescope constructed by Paul Shuch of the SETI League.

The same situation might obtain for SETI. It's possible that some civilizations are routinely aiming strong beacon signals in our direction – either radio or light – "pinging" us intermittently. The professional SETI experiments, which systematically scrutinize small areas of the sky, could easily miss such occasional "signal flashes." But large numbers of amateurs could effectively monitor the entire sky (albeit with lower sensitivity), and possibly find such high-powered pings.

The SETI League, in New Jersey, runs the most ambitious of the amateur SETI experiments. More than 100 volunteers in nearly two dozen countries are engaged in the League's Project Argus, a radio SETI experiment using backyard satellite dishes. The participants rig up their own equipment (the cost is hundreds to thousands of dollars, depending on the technical expertise of the amateur), and coordinate their search via the League. The diameter of these dishes is typically 3 to 5 meters, so that they individually survey a chunk of sky roughly 4 degrees across, a patch the size of 50 full Moons. And like most SETI experiments, they generally tune their receivers to the so-called "water hole" near 1,400–1,700 MHz.

Project Argus is hoping to recruit five thousand observers. You can learn more about this project at the SETI League's website, www.setileague.org. There you'll also find links to organizations that supply hardware for amateur radio astronomy telescopes that you can build for studying supernovae, active galaxies, and some natural, Solar System radio emitters.

Amateur optical SETI – looking for very bright, but very short flashes of light from nearby star systems – is still in its infancy, although it's interesting to note that a SETI amateur, Stuart Kingsley (in Columbus, Ohio) has long been both a pioneer in the field and one of optical SETI's greatest champions. Companies that supply equipment for amateur astronomy may soon be building devices to attach to a mirror or lens telescope; equipment that would enable amateurs to search for flashes from the sky.

Finally, the easiest way to get involved in actually doing SETI research is to fit out your home computer with the University of California at Berkeley's popular screen saver, SETI@home (see sidebar).

existing telescopes – huge, costly antennas that have been constructed for conventional astronomy research. For example, Project Phoenix – the most sophisticated SETI experiment to date – has run on antennas in Australia, West Virginia, Puerto Rico, and Britain. Borrowing someone else's telescope is less costly than erecting your own, but the disadvantage is that you only get a limited amount of listening time.

One way out of this bind is to "piggyback" on a telescope while it's busy with traditional studies of galaxies, pulsars, or whatever. This is somewhat like hitch-hiking – it saves you the expense of a car, but on the other hand you can't steer, and you don't always proceed along the most direct route to your destination. Two major radio SETI experiments today are in piggyback

Frank Drake in West Virginia. Drake conducted the first modern SETI experiment in 1960.

mode: SERENDIP IV and Southern SERENDIP (SERENDIP is an ingenious, if somewhat tortured acronym for the Search for Extraterrestrial Radio Emissions from Nearby Developed Intelligent Populations). SERENDIP IV takes advantage of a second receiver on the economy-size Arecibo telescope in Puerto Rico, and Southern SERENDIP does something similar using the 64-meter Parkes Radio Telescope located in the sheep country west of Sydney, Australia.

Hitching a ride on someone else's telescope is an excellent way to collect lots of data. The fact that it's the radio astronomers, and not the SETI observers, who determine where to point the antenna may not be a crippling disadvantage. After all, until we find a signal, we really can't be sure where in the sky the aliens are situated. In the interim, you can argue that one spot of sky is just as good as any other. To put it a different way, if you don't have any idea where the ducks are, shooting randomly into the air seems as good a strategy as any. But not everyone agrees. As we've noted, Frank Drake spent his limited telescope time checking out Sun-like star systems, not just any old patch of celestial acreage. This seemed like a reasonable plan at the time (after all, the Sun has planets, including one that's known to be frothing with life). It seems an even more reasonable strategy now. Since 1995, astronomers have been discovering worlds around other stars, stars that are often very much like the Sun in bulk and brightness. So most researchers concur that it makes sense to direct our attention to the vicinities of Sun-like stars, precisely as Drake had supposed.

Several SETI projects do this. Project Phoenix, run by the SETI Institute in California, is busy checking out roughly one thousand Sun-like star systems in our cosmic backyard. The "backyard" refers to the part of our Galaxy within 200 light-years. (200 light-years is about 2 thousand trillion kilometers, which may sound like an impressively large backyard, but this distance is only 0.2 percent of the Galaxy's diameter.)

The digital receivers that gather the data crunched by the SETI@home program.

Pointing the telescope is one thing, but where on the dial should we listen? To what frequency do we tune the receivers? Astronomers have learned that the Cosmos is particularly quiet in the so-called microwave region of the spectrum. This lies between about 1 and 50 gigahertz (GHz), a range of frequencies used for such practical purposes as satellite communication and heating leftovers. At the low end of the microwave range are a couple of frequencies of particular interest: 1.4 and 1.7 GHz. These are frequencies at which some common components of the thin gas between the stars emit natural cosmic noise: hydrogen (H) and the hydroxyl radical (OH). Since H and OH combine to form water – which is usually described as the indispensable ingredient for life – the radio dial between 1.4 and 1.7 GHz is often called the "water hole." Earthly radio astronomers spend a lot of time tuned to these frequencies, and you can be sure that extraterrestrial astronomers do too. Since we can presume that everyone in the Universe has their receivers marked at both 1.4 and 1.7 GHz, the water hole is an obviously promising band for interstellar communication. Ever since Project Ozma, it has been where SETI researchers tune in most often.

The 64-m radio telescope at the Parkes Observatory, in New South Wales, Australia.

The SETI Institute, in Mountain View, California, is home to those looking for life in the Universe.

But *what* do they hope to tune in to? Complex signals filled with information about some alien culture? Obvious mathematical series used as an interstellar Morse code? Musical tones designed to simultaneously amuse and inform? Alien TV commercials advertising fast cars and dental care products?

We don't know and we *won't* know until and unless we pick up ET's broadcast. Meanwhile, SETI researchers simply program their receivers to look for signals that are confined to a small part of the dial – a type of signal that only purpose-built transmitters can make. They'll search for the actual content – or message – later. In other words, their strategy is to first discover if the aliens are on the air, and once that's done then they'll build

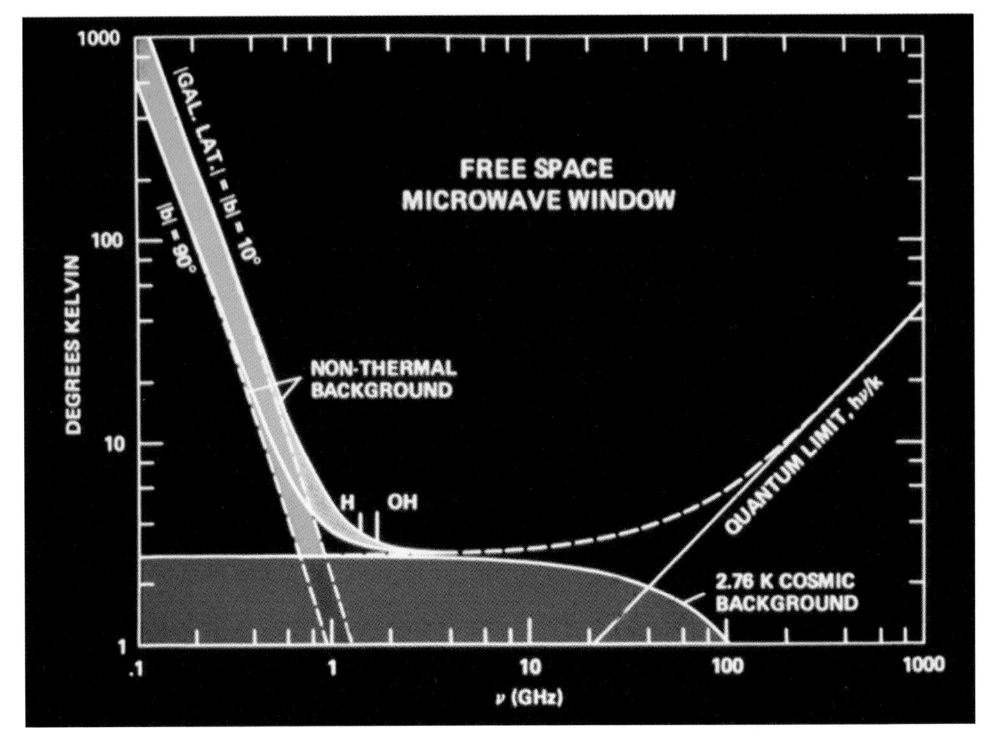

This chart, which shows how much natural cosmic static exists at various frequencies on the radio dial, has an obvious quiet spot that includes where the letters H and OH are. These letters correspond to the frequencies at which hydrogen (H) and the hydroxyl (OH) radical emit natural radio noise. The frequencies in between are known as the "water hole".

the equipment needed to learn if they're sending us their algebra, their art, or their marketing hype.

Any signals?

It's been more than four decades since Project Ozma first tried to eavesdrop on alien broadcasts. Have SETI researchers ever heard anything?

As it turns out, they hear signals all the time. Regrettably, none of these has ever been proven to be extraterrestrial; they all seem to be merely interference from transmitters on Earth or from telecommunication satellites that circle our planet. But are the SETI scientists "close"? Given the big telescopes they now train on the skies, can we expect that they'll pick up a tell-tale signal soon?

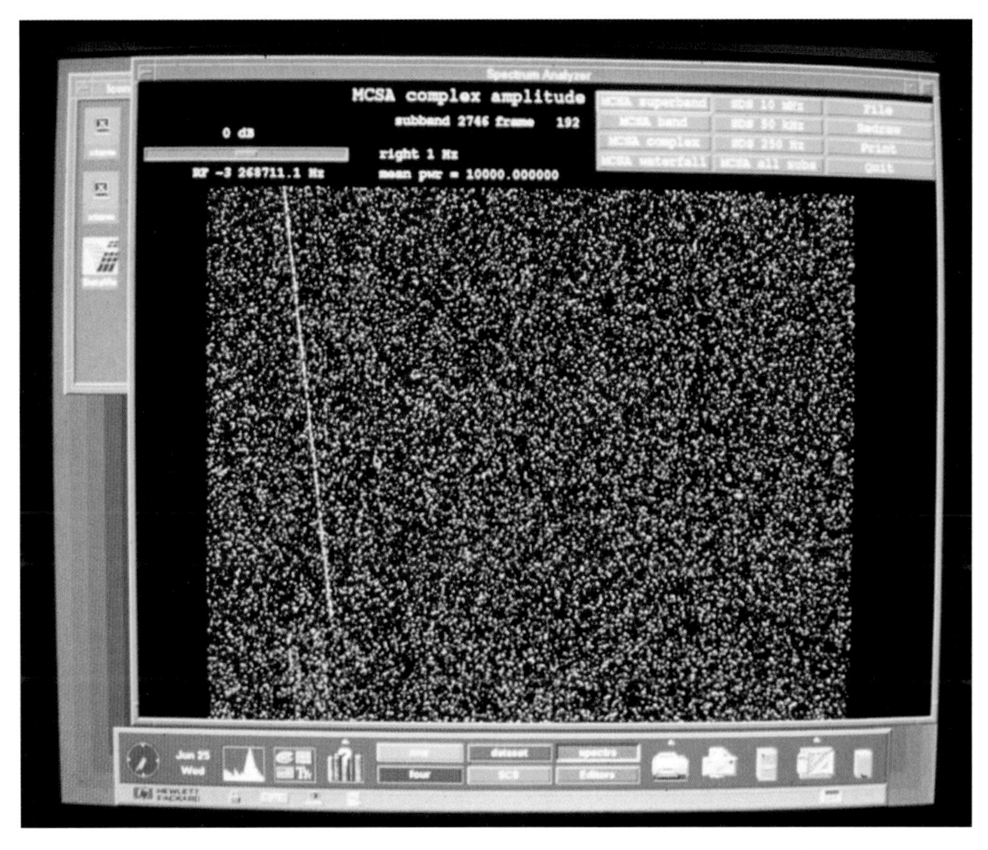

This signal is "extraterrestrial." It is coming from Pioneer 10, which is travelling out of the Solar System. This spacecraft, launched in 1972, is now more than 1.5 times as far as Pluto. You can see what a distinctive pattern a transmitter makes against the background noise.

This is a bit like asking James Cook, sailing somewhere amid the vast, salty wastes of the South Pacific, whether he's "close" to discovering a new island. He can't possibly know. On the other hand, the longer he searches, and the more water that passes under his keel, the greater the chance of discovery.

There's no doubt that SETI researchers are covering far more of the celestial seas in their searches today than ever before. Furthermore, the sensitivity of their experiments is also increasing. Project Phoenix, for example, could detect an incoming signal even if the radio energy collected by the telescope was a trillion times less than a falling pillow feather. New instruments that are now being built will speed the search (see Chapter 7). It's

Signals that tease

Strictly speaking, there's no such thing as a "close call" for SETI. Either a signal is truly from an extraterrestrial intelligence, or it's not.

In reality, SETI researchers frequently pick up signals that mimic the appearance of an other-worldly transmission. After all, they're using monstrously large antennas for their radio searches, and observatory-grade optical telescopes in the hunt for alien light pulses. These instruments are frighteningly sensitive, so it's hardly surprising that the researchers often trip over natural phenomena or the radio clutter of our own planet. (The signals produced by dead stars known as pulsars are so regular that when they were first detected, the scientists jokingly referred to them as LGMs, or Little Green Men.) Such non-alien signals can be hard to recognize, and SETI has occasionally found signals that, at least for a while, looked like the real deal.

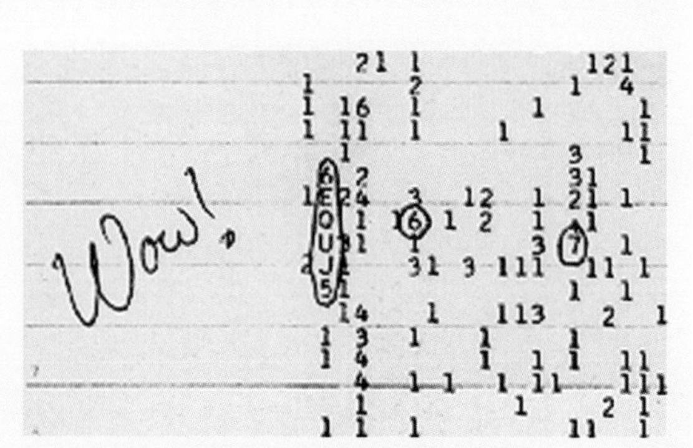

The "Wow" signal. Interference? Or was ET briefly on the line . . .

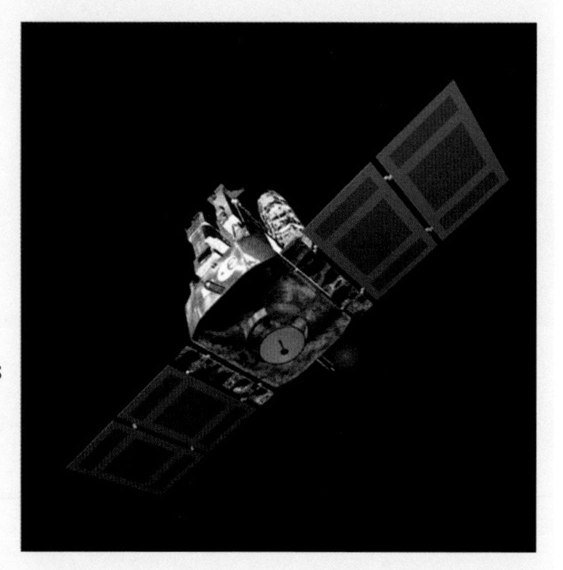

SOHO was launched in December 1995 and operational by 1996. It is nearly 8 meters in size.

Perhaps the most famous such find was made more than two dozen years ago with the "Big Ear" radio telescope at the Ohio State Radio Observatory. This antenna, which has since been demolished, consisted of a large parabolic reflector on one side of a field and a tiltable, flat reflector on the other. For 25 years it automatically recorded and printed out incoming cosmic static in its patient search for extraterrestrial broadcasts. On August 17, 1977, astronomer Jerry Ehman was reviewing the computer printout of a recent observing run, and found an unexpectedly strong signal. Indeed, Ehman was so impressed that he wrote "Wow" on the chart paper next to the signal. Was this ET calling?

Probably not. The signal was never found again. A few minutes after the initial detection, a second beam of the "Big Ear" automatically passed over the same patch of sky. No signal was detected. More recent efforts to detect the "Wow" signal, using the Very Large Array in New Mexico and a 26-meter antenna in Australia, have also come up empty-handed, despite the fact that these searches were considerably more sensitive than the original Ohio State scan. The "Wow" signal has become a staple of both science fiction and Internet chat groups, where interested amateurs moan that scientists are ignoring this clear evidence for extraterrestrials. But a signal found only once is not sufficient to convince SETI researchers that they've discovered the aliens.

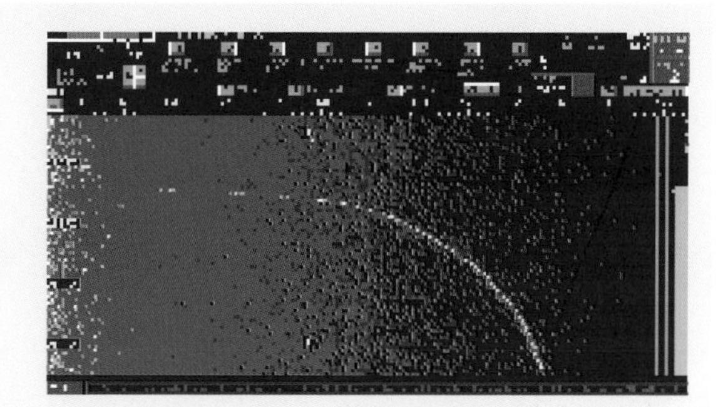

The "hacked" EQ Pegasi observation.

A similar tale can be told for every other "interesting signal" so far detected, including more than a hundred picked up in a search, known as Project META, conducted by the Planetary Society. Follow-up observations have failed to confirm these intriguing finds. In 1997, Project Phoenix, using the 43-meter radio telescope in West Virginia, encountered a faint radio whine that had all the hallmarks of an extraterrestrial transmission. As you can imagine, the blood pressure of the Phoenix researchers rose significantly. It took most of a day to figure out that the signal was really being beamed from the SOHO solar research satellite, located more than a million kilometers from Earth.

In 1998, an Internet website claimed that a SETI signal coming from the star EQ Pegasi had been found by a British engineer. This, too, caused some short-lived excitement, but within a week it was realized that the signal was simply a hoax. Even optical SETI experiments routinely get unanticipated "flashes" from star systems in their sights. But so far, none of them have been repeated.

The bottom line is simple. Any extraterrestrial signal worth its salt, or at least worth mention in the reputable news media, must be persistent enough to survive repeated efforts to confirm it. That's the way of serious research. Of course it's possible that the "Wow" signal or one of these other

SETI scientists Peter Boyce, Jill Tarter, and Peter Backus using the Parkes telescope as part of Project Phoenix.

teasers was truly an extraterrestrial civilization calling – and when the astronomers tried to find it a second time, the aliens had switched off the transmitter. But scientists are a skeptical lot. If you want to claim something as important as the detection of a society around another star, the evidence had better be solid and repeatable.

possible that one day soon we'll find out that Earth is not the only planet occupied by thinking beings.

Flashing lights

Radio broadcasts aren't the only way to signal your presence across interstellar space. Ordinary light could do the trick, too.

This sounds reasonable: after all, we've long used light as a signaling scheme on Earth (think of ship telegraphs). But there's an apparent problem: wouldn't luminous signals from distant solar systems be drowned out by starlight? Stars produce rather little radio emission, and therefore wouldn't interfere with alien radio broadcasts. But Sun-like stars pump out more than 100 trillion trillion watts of blinding visible light!

You might think that it would be difficult to construct an interstellar beacon intense enough to outshine that. But in fact it's not so hard. This is because stars radiate their light in all directions. What we (or the aliens) could do is to fit a mirror or lens on our light source to focus it into a beam that's only aimed in the direction of intended recipients.

The Harvard Optical SETI project on the 1.5-m Oak Ridge Observatory telescope in Massachusetts.

For example, suppose we wanted to ping a star system 1,000 light-years away, using a high-powered laser as a transmitter. If we focus the laser with a 1-meter diameter mirror or lens, the beam will barely spread to the size of our Solar System (measured to the orbit of Saturn) after traveling 1,000 light-years. That's sufficient spread to flood with light any inhabited planets around the target star, and still narrow enough that we haven't wasted our light on far-out, lifeless worlds. A focused beam this size illuminates a patch on the heavens that's only about a hundredth of a trillionth of the entire sky. So if our laser can manage a power output greater than a trillion watts, it can outshine a Sun-like star! Lasers capable of reaching this power level – at least for very, very short bursts – are within *our* technical capabilities today, even though the first laser was built only 45 years ago. Surely there will be sophisticated societies out there for whom the construction of such flashy hardware is a trivial effort. If we can do it, they can do it. So it seems to make eminently good sense to search for pulsing lights coming from the directions of neighborhood stars.

This is exactly what's being done at several observatories in the USA. Light detectors able to register flashes as short as a billionth of a second have

Optical SETI at Lick Observatory. The white box behind the Nickel 1-m telescope was built by Shelley Wright, a student at the University of California, to search for light pulses only a billionth of a second long.

SETI@home

Radio SETI experiments look for simple signals: either a narrow-band monotonic tone (which, if you played it through your audio system, would sound like an endlessly-sustained, pure musical note), or a slowly pulsing version of the same thing (which would sound like a note going on and off). These types of signals don't convey a lot of information, but would at least tell you that someone's transmitter was "on." The fact that such transmissions are confined to one spot on the radio dial, which is the definition of "narrow band," makes them quite unlike the natural crackle of quasars, pulsars, or other cosmic objects. Because a narrow-band signal concentrates radio energy into a small range of frequencies, it would stand out better against the general background static of the universe. It would be more economical for the aliens to transmit, and easier for us to receive: a good idea all around, and a great way to get attention.

The digital receiver used for Project Phoenix can simultaneously monitor incoming cosmic static from 58 million channels.

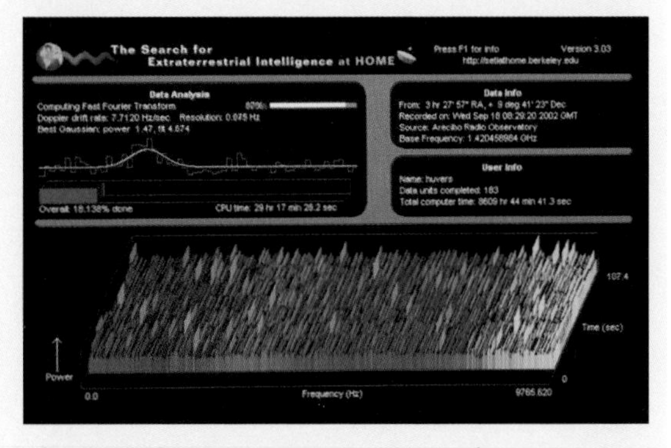

The SETI@home screen saver at work.

In searching for such signals, today's SETI experiments often pore through tens or even hundreds of millions of channels at a time. Consequently, the rate at which they collect data can be enormous – enough to fill a CD every few seconds. It's because of this torrent of numbers that experiments such as Project Phoenix require sophisticated, purpose-built computing hardware to keep up.

But not all SETI experiments need to process their data right away. For example, a project such as SERENDIP IV, which is piggybacked on the Arecibo telescope, has less need to immediately swallow the flood of incoming bits. This is because such piggyback experiments don't control where the telescope is aimed – every few seconds, they snatch data from a different piece of sky. So even if a signal is found right away, it can't be checked out with a second observation until, by chance, the telescope returns to the patch of sky in which it was first discovered. With a piggyback scheme, the check-out time can be months or even a few years. So signal processing needn't be "real time."

This has encouraged the SETI researchers at the University of California in Berkeley to use a very clever scheme for processing some of the SERENDIP IV data. They've broken up a small percentage of their data into 250 kilobyte chunks, and now distribute it via the Internet to folks who have downloaded a free screen saver onto their home computers. The screen saver, called SETI@home, runs on your computer whenever you take a kitchen or bathroom break. The first thing it does is download onto the hard drive one so-called "work-unit" – a chunk of telescope data plus about 100 kilobytes of additional information. The computer carefully analyzes these data for signals over the course of a few days or weeks (depending on how long you leave your machine idle). Once analyzed, the results are sent via the Internet back to the project headquarters in Berkeley, and another work-unit is downloaded onto your computer.

Because more than four million folks have downloaded SETI@home, each piece of data can be very carefully scrutinized. The screen saver looks for signals of various (narrow) bandwidths and pulse rates, and signals that might drift slowly up or down the dial. The enormous computational power of millions of machines means that even when looking for the simplest types of signals, SETI@home can dig far deeper into the background static than is usually the case.

Of course, there's no instant gratification with SETI@home. Any signals uncovered by your trusty home computer will need to be compared to other observations of the same bit of sky, generally made months apart. Most (and possibly all) are destined to be ultimately identified as terrestrial interference, caused by satellites, radar, and other earthly noise makers. Indeed, in March, 2003, the SETI@home scientists carefully re-observed 166 of the best candidate signals (among many millions) bagged during the first few years of SETI@home analysis. None of these was confirmed to be a true, extraterrestrial signal. They were all earthly interference. But future work may yet prove that a faint whine discovered with someone's home computer is from a world other than ours. If it's your computer that finds that signal, then you will be the first person in all history to prove the existence of intelligent beings in the vast, inky spaces of the Cosmos.

What listening for aliens is really like

In the movies, finding alien signals on the radio is pretty straightforward.

Silver-screen SETI researchers usually make their discoveries while sitting around looking mildly bored, monitoring incoming cosmic static on a handy loudspeaker or with a pair of earphones. When a signal turns up (and in the movies, it *always* turns up), the researcher is alerted by a weird-sounding noise. Surprised and wide-eyed, he then makes a quick check on a graphic display – depicting a wave form reminiscent of the Himalayas seen in profile – and then breathlessly spills the news to colleagues, the government, and the world at large.

One of the authors at work at the Arecibo Observatory. This is the control console for Project Phoenix.

Signal detection in real SETI experiments isn't quite so simple. Since these experiments typically use receivers with many tens of millions of channels, it would take a room stuffed with loudspeakers or a truck full of earphones to monitor them all – not to mention the long-term attention of millions of listeners. In real life, computers do the "listening" – and only bother the SETI researcher when they find something that truly seems to be an extraterrestrial transmission. Besides simplifying the process and reducing the boredom, the use of computers to search for possible broadcasts has other advantages. When it comes to fishing weak signals from the gush of natural static, computers are superior to people. In addition, machines can easily recognize signals that pulse on and off, or that slowly move up or down the radio dial.

The swelling tide of radio interference caused by our own civilization has to be sorted out in any SETI experiment, and for this, too, computers are better. They start by checking all incoming signals against a database of known terrestrial interference. This allows many of these annoying local noise sources to be automatically rejected. Signals that don't drift slowly in frequency are thrown out, too. The rotation of our planet ensures that any extraterrestrial signals will change in frequency during the course of only a few seconds. All such checks are made automatically by the patient computers.

The Jodrell Bank telescope in Cheshire, England, used to check out signals found at Arecibo by Project Phoenix.

Even so, there are still many signals that pass such automatic checks, and need to be examined in greater detail. The most sensitive SETI experiments do this by enlisting the help of a second radio telescope. When Project Phoenix is observing nearby stars with the Arecibo dish in Puerto Rico, the most promising signals are followed up by the 76-meter Lovell radio telescope at Jodrell Bank, England. The use of two antennas enormously reduces the "false alarm" rate.

Before SETI researchers can confidently claim that they've found ET, several days of testing and checking will be required, and scientists at a third telescope will also be asked to confirm the signal. This independent check is essential to rule out subtle software bugs, unknown terrestrial interference, or a college prank. Clearly, the real process of finding a signal is a lot slower than what happens on the screen, but then again cinema audiences would probably balk at a week-long movie.

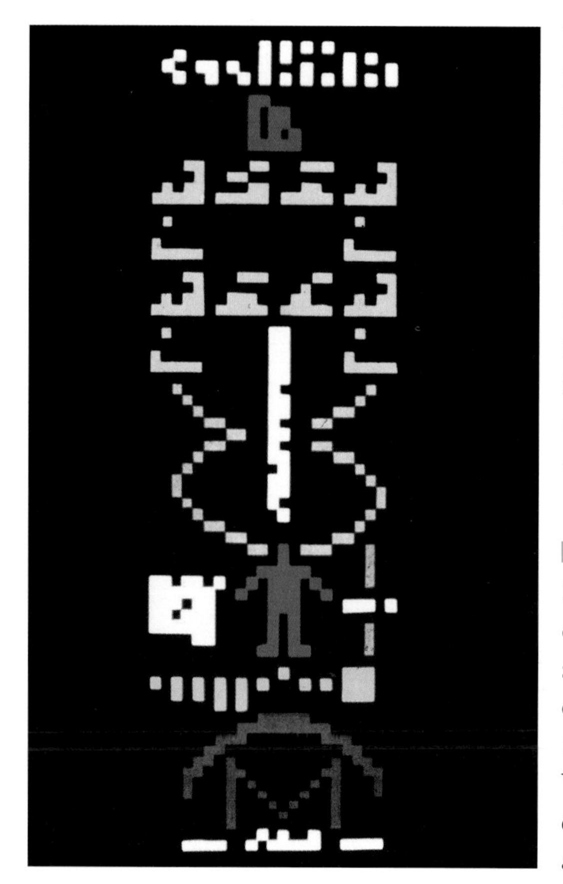

This binary message was sent to the stars. Could you decode it?

been fitted to conventional mirror-and-lens telescopes. In fact, multiple detectors are normally used so that random flashes from cosmic rays or natural radioactivity can be recognized and tossed out. Several thousand stars have already been checked.

What's the bottom line? Just as for the radio SETI experiments, no clear-cut light signals from the sky have been received. But it's still early days for such efforts. Proof of the existence of technically competent aliens could arrive tomorrow on a light-beam.

Do we broadcast?

So far, all SETI efforts have been passive. This doesn't refer to the docile nature of the researchers, but rather to the fact that they've only been listening or looking. No serious broadcasts intended to provoke replies are being blasted into the Cosmos by SETI scientists. Of course, as already noted, humans are unintentionally transmitting the most refined aspects of our culture into space via radio, television, and military radars. But readers with a technical bent are probably aware that within a decade or two, far fewer such transmitters will be spouting entertainment to possibly appreciative neighbors. Most of us will be receiving programs via fiber optic cables that snake into our homes, or from direct satellite broadcasts.

The aliens, we can safely presume, have already made this technical switch. The amount of radio (or light) energy leaking from their worlds might be quite modest. If this is the case, then we will only find the aliens if they are making deliberate transmissions. But if *we're* not broadcasting, why should they? What if everyone is listening?

Doesn't this suggest that we should be on the air, sending signals to the stars?

Probably not. Indeed, there are several reasons why we don't broadcast. Perhaps the most compelling of these is the fact that our society has just arrived on the technological scene. Powerful radio transmitters and lasers are twentieth-century inventions. For an advanced civilization elsewhere, this kind of stuff is more than old hat – it's ancient achievement, in the way that fire and the wheel are old (but still useful) inventions to us. Since humans are the newest players on the technical stage, it certainly makes sense that we would choose to listen first, and save the mouthing-off for later.

And incidentally, if you are puzzled by why an alien society would transmit at all, several motivations come to mind. Perhaps they need to keep in touch with space colonies, or use hefty radars to search for comets on a collision course with their planetary home. And, of course, some of these societies might actually *want* to get the attention of neighbors, and possibly instruct them in the finer points of cosmic existence.

It's worth noting that a few SETI researchers have argued that even a modest transmitting project from Earth would have a practical benefit. Broadcasting would help us to figure out the best way to *listen*. Perhaps that's true, but so far neither governments nor individuals have been willing to put up the monies necessary to set up a long-term, effective cosmic equivalent of the BBC. Our efforts to get in touch with intelligent cosmic company are limited to passive experiments to find signals.

All these experiments have yet to find an alien transmitter, and it's time to consider in greater detail whether the chances for success are good – or otherwise.

CHAPTER 6

The Drake Equation

Day after day, SETI researchers point their antennas and mirrors skyward in hopes of finding evidence of intelligence. But what are the chances that their gamble will pay off?

It's hardly an unreasonable question. If you've signed on for a voyage of discovery, it's always worthwhile to gauge the probability of success. So it's not surprising that very soon after Project Ozma, scientists were already trying to reckon the odds of making a SETI detection.

The Arecibo telescope is used round the clock by astronomers. About five percent of the time is allocated to SETI.

Frank Drake demonstrates his equation.

The year they first did so was 1961. Frank Drake was co-hosting the first-ever SETI conference in Green Bank, West Virginia – at the radio observatory where he had conducted his pioneering search only a year earlier. He wanted some way to organize the discussion, some sort of agenda for the conference. Everyone had said that finding a transmitting society hidden in the black depths of the Universe was akin to finding a needle in a haystack. Well, of course that was true. But Drake realized that the degree of difficulty could only be estimated if you knew how many needles were hidden in the hay. So he organized the conference around a simple equation that would compute how needle-rich the Galaxy might be.

Drake's Green Bank conference lasted three days. But the formula that was intended to serve as an agenda is still around. Known as the Drake Equation, you can find it adorning the pages of just about any textbook that deals with the matter of life in space.

Multiple generations of stars are being born in the Tarantula Nebula.

The equation explained

Most people aren't very keen on equations. Fortunately, the Drake Equation is so simple that even the math-phobic reader can understand it (after which you can impress your friends with new-found knowledge).

The intent of the equation is to compute *N*, the number of civilizations in our Galaxy that have sent signals reaching Earth now. We don't care whether they're transmitting radio waves or using big lasers to flash our world, since SETI experiments look for both types of signals. The idea is simply to calculate how many societies are sending detectable signals. Obviously, if this number is high – for example, if there are millions of signaling civilizations – then SETI researchers should expect success fairly soon. On the other hand, if N is very small – say two or three – then it may take a very long time to discover cosmic company. N is the number of needles in the galactic haystack.

Conceptually, the Drake Equation can be understood by considering an analogy. Suppose you wish to compute how many police dogs, worldwide, are out and about. Admittedly, this is a peculiar thing to want to do, but it will serve to illustrate the logic. The simple way to proceed is to calculate how many dogs each year are indoctrinated into the K-9 force (the rate of recruits), and then multiply this rate by the number of years the average hound stays in service. For example, if 10 thousand dogs are sworn into duty each year, and the average dog spends four years fetching fugitives, then there are about 40 thousand police canines on the beat.

In other words, the number $N =$ rate $\times$ lifetime, in which "lifetime" refers to how many years the average dog serves, not how many years he lives. We can probably guess the lifetime – after all, it's likely to be some number between one and ten years, depending on how dangerous the neighborhood is. It might be a bit harder to come up with a number for the rate at which police pooches are indoctrinated. To help us compute this, we could imagine breaking the rate up into yet simpler terms. For example:

Rate of dogs inducted into service =

- the annual birth-rate of *all* dogs, worldwide (R), *times*
- the fraction of dogs that are of a breed suitable to become police dogs (f_s), *times*
- the fraction of suitable dogs that actually enter K-9 training (f_t), *times*
- the fraction of K-9 recruits that pass training (f_p).

In this way, we've factored the rate at which dogs enter the force into a string of simpler terms. Note that we've also given symbols for each of these terms. Having done so, we can write our equation to estimate the number of police dogs, $N = R\, f_s\, f_t\, f_p\, L$, in which L is the average lifetime on the force.

Now that we've tediously established the very simple logic of this approach, we can dispense with the dogs, and lay out the terms of Drake's equation for estimating N, the number of communicative civilizations. It, too, is computed as a rate times a lifetime. The rate is now the number of signaling societies that arise each year in the Galaxy. The lifetime L is the average number of years such a technologically adept society survives – in other words, the length of time they "stay on the air."

The rate is computed as the product of the following terms:

Rate of appearance of signaling societies =

- the annual birth-rate of stars that could support habitable planets (R_*), *times*
- the fraction of such stars that have planets (f_p), *times*
- the average number of habitable planets and moons per solar system (n_e), *times*
- the fraction of such worlds on which life actually springs up (f_L), *times*
- the fraction of worlds with life on which intelligence evolves (f_i), *times*
- the fraction of worlds having intelligence that produce technical civilizations (f_c).

So the full-blown, adult version of the Drake Equation is

$$N = R_* \, f_p \, n_e \, f_L \, f_i \, f_c \, L.$$

Every term of the equation was the subject of discussion at the 1961 Green Bank conference, and the attending scientists did their best to come up with numbers to hang on each one. But while it's not too hard to estimate the galactic birth-rate of stars (after all, that's straightforward astronomy), it's much more of a stretch to place some value on the other terms to the right of the equation. For example, how many planets with biology will produce creatures with intelligence? Obviously it happened here, but maybe *Homo sapiens* was just an improbable accident (and from the point of view of beef cattle or tuna, an unfortunate one).

You can find estimates for N in the professional literature that range from one (we're the only world with sophisticated inhabitants in the entire Galaxy!) to millions (we've got lots of company). It's likely that we'll only get a better idea of the true value of N once we've found other worlds with intelligent life, assuming that we eventually do.

Despite this mammoth uncertainty, the Drake Equation has demonstrated more staying power than red wine on a white carpet. It's the point of departure for any SETI discussion. In the next section, we'll describe the terms in somewhat greater detail, and indicate what modern research says about their value.

The equation quantified

In 1961, the only term of Frank Drake's formulation that could actually be estimated with good reliability was R_*, the rate at which suitable stars are born in the Milky Way. An easy way to make your own estimate is to note that there are a few hundred billion stars in the Galaxy, and the Galaxy itself is roughly 13 billion years old (large galaxies were formed a billion or so years after the Big Bang). Dividing these two numbers, we see that, if the Milky Way has always spawned stars at a constant rate, then a few dozen suns are born each year. As we've discussed before, some of those suns – particularly the large ones – aren't very suitable for hosting habitable planets. But at least half of all stars are, and maybe more. So we can say that R_* is likely to be at least 10 stars per year.

The value of the second term, f_p, would have been a complete guess in 1961. At that time, no planets had ever been detected beyond our own Solar System. However, as we described in Chapter 1, worlds around other stars are now cropping up faster than corn in August. All of these new-found worlds are giant-size, which means they're probably not the most attractive for begetting life. But the reason that our catch is limited to the bigger planets is that our net is too coarse. Small worlds, similar in size to Earth, must surely be out there, but proving that's true won't happen until we've launched some new telescopes into space later in this decade. In the meantime, all we can say is that at least one star in ten has planets (f_p is 0.1 or more).

The value of n_e, the number of worlds per planetary system that could be habitable, is less certain. In our own Solar System, we have Earth, Mars, several moons of Jupiter, and one Saturnian satellite as candidate worlds for life. So a value of 2 or 3 for n_e might not be a bad guess. (Of course, based on one stellar system, it's no more than a guess!)

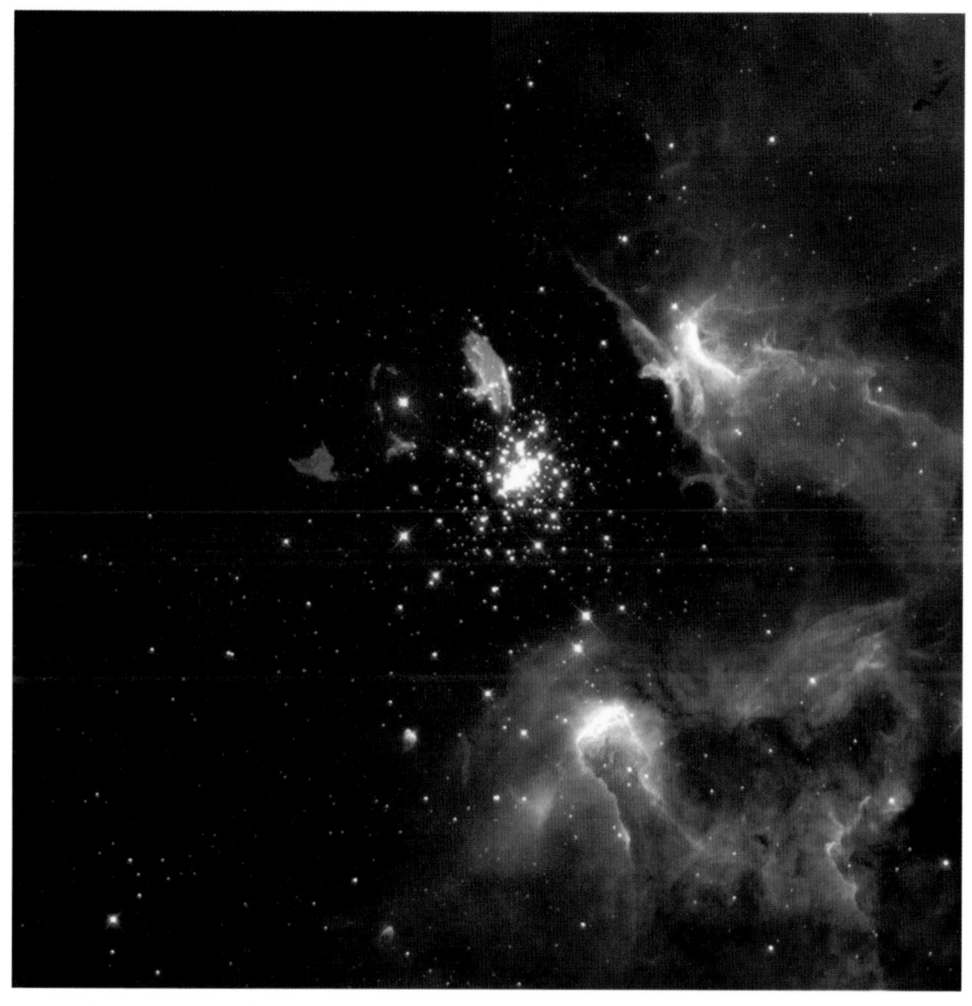

A star cluster emerges from its nebula.

Now we come to f_L, the fraction of those candidate worlds where life arises. Whether this is very small or close to 1 depends on whether we think that life is near-miraculous (Earth is lucky) or a natural affliction that will appear on just about every planet with a bit of liquid water. As we mentioned in Chapter 2, life began on Earth nearly as soon as it could; once the heavy bombardment of rocks left over from the Solar System's birth eased up, biology was underway. That suggests that biology is a simple project for nature, and that f_L is probably near 1. On the other hand, we'd

Our civilization eventually developed the means to communicate. Will aliens also build radio transmitters?

be a lot more confident of this if we found first-rate evidence for alien life (even dumb microbes) on Mars or some of those appealing moons in the outer Solar System.

As we move farther to the right in the Drake Equation, the terms become still more speculative. What fraction f_i of worlds with life will eventually witness the appearance of intelligent life? We've discussed the evolutionary pressures in a competitive environment that might prod complex animals towards intelligence. On the other hand, many evolutionary biologists are not so sure that big brains – which are a major metabolic expense – are inevitable or even likely. This is a term that optimists will assume is near to 1, and the less enthusiastic will peg closer to 0.

Even if crafty creatures appear, will they ultimately develop a technologically competent civilization and hence be able to communicate? There have been (and still are) many human societies that have little interest in the idea of ongoing, technological progress. The ancient Egyptian civilization, for example, hardly changed for thousands of years. You may think it's completely natural that every generation will have better science and technology than its predecessor. But there's no guarantee that our cosmic company agrees. On the other hand, understanding science is remarkably useful and beneficial. Even if it's only developed by a small subset of any species' population, it's likely to spread widely. This is our experience on Earth, and if we assume, once again, that we're typical, then f_c is probably close to 1.

Finally, we come to the last term in Drake's enticing equation: L, the lifetime of a technological civilization. Like f_c, this factor is completely dependent on "soft science" – sociology, rather than astronomy or biology. In addition, it has been forcefully argued that L is the term that we know, and perhaps *can* know, least about.

These Thor and Blue Streak rockets started life as ballistic missiles and then were put to more peaceful space exploration uses.

What is our experience on Earth? We've been a signaling society for about 70 years. But shortly after we invented radio, we developed nuclear weapons. Perhaps this means that we – and most aliens – are doomed to self-destruction within a century or two after going "on the air." If so, if L is only a few hundred years, then our chances of finding another civilization are small, and our future is grim. The very technology that allows an intelligent species to get in touch might also be its undoing.

While this is certainly possible, there's another, less morbid way to look at the situation. Within a century after the invention of radio, *Homo sapiens* was on the Moon. In another century, it's quite likely that human colonists will have set up shop not just on our neighbor satellite, but also on Mars,

The International Space Station is a first step in becoming independent of planet Earth.

and possibly in large, artificial habitats in orbit around the Earth. Our expansion into the Solar System will have begun.

Once this happens, it's safe to say that we will have inoculated ourselves against self-destruction. Even if massive wars break out on Earth, there would still be the colonists elsewhere to pick up the pieces. The situation would be somewhat analogous to the horrific plagues that swept Europe in medieval times. While locally devastating (one in three people died), the plagues actually had rather little effect on the total population of humans, most of whom lived in places *other* than Europe. So perhaps the depressing scenario touted by the pessimists is only part of a bigger, more optimistic picture – a society invents radio, and shortly thereafter develops the capability for self-annihilation. But the rockets that can be used to deliver weapons can also deliver inhabitants to other, nearby worlds. Yes, technological societies will endure a dangerous century or two when they are vulnerable to obliteration. But at least some of them (one hopes) will make it past this intimidating bottleneck, and become sufficiently spread out to ensure that their species broadcasts its presence for thousands or millions of years. If so, then L might be very large indeed.

In this discussion, the astute reader will note that we've deliberately avoided giving "official" values to the terms of the Drake Equation. This is because there *are no* official values! In general, you can plug in whatever numbers appeal to your good sense and personal esthetics (excepting R_* and f_p, for which astronomy has provided meaningful constraints). As mentioned, SETI scientists who've wrestled long and hard with the equation have come up with a wide range of estimates for N, ranging from zero to tens of millions. Drake himself reckons that N is about 10,000. Ten thousand communicating civilizations in our Galaxy. The importance of such estimates is that – to the extent you believe them – they tell you how many stars must be searched in our experiments, and roughly how far we are from any cosmic companions.

Limitations and other possibilities

While the Drake Equation is still – after four decades – an extremely fruitful way to consider the likelihood of a SETI detection, it has some limitations. And several of these might be important.

One point is that the calculation is made for our galaxy only. What about the 100 billion or so *other* galaxies? Of course these might have inhabitants, too. Who would deny the bulk of the Cosmos the obvious blessing of thinking beings? But most SETI experiments limit their searches to our own galactic home. This is merely being practical; other galaxies are too far away. Even Andromeda, the nearest large galaxy, is twenty times farther than the most distant stars of the Milky Way. Other big galaxies are farther still. Not only do signals weaken with distance, but the incentive to broadcast does as well. After all, how interesting is it to ask questions when replies might take millions of years?

Perhaps a more serious concern with the Drake Equation is its anthropocentrism. The formula is based on the assumption that other life forms arise under conditions similar to those on Earth – a nice planetary home that spawns life, and (eventually) intelligent creatures. Each society arises independently, and each stays close to the star of its birth.

How far are the aliens?

How many light-years is it to the nearest alien civilization? In 1960, when Frank Drake made the first modern effort to eavesdrop on radio signals from ET, he trained his antenna on two relatively close stars, Epsilon Eridani and Tau Ceti; respectively 10 and 12 light-years from Earth. He picked these neighborhood stars for several reasons. For one thing, they are similar in size and brightness to the Sun, the type of star most likely to have planets suitable for the dirty chemistry we call life. But of great importance, they are nearby, and signals from close-in transmitters are likely to be stronger. They will suffer less from the inevitable dilution of distance.

In addition, it would be more interesting to find aliens in our neighborhood, as opposed to halfway across the Galaxy. If the aliens are *really* close – say less than 100 light-years – then two-way communication, while admittedly slow, would at least be thinkable.

Carl Sagan.

Drake didn't hear anything from these solar siblings. Nor have subsequent SETI searches, including the SETI Institute's Project Phoenix, a scrutiny of a thousand nearby stars. What this tells us is that the Universe is not laced with strong, persistent signals from advanced societies. Apparently, evidence for extraterrestrials is not trivial to find, and the signals we seek are likely to be weak.

Of course, this wasn't known in 1960. After all, it was possible that extraterrestrial civilizations were so rampant that even the stars next door might house aliens. But the SETI experiments of the last decades have shown that's not the case. This suggests that our local galactic neighborhood is not chock-a-block with alien societies.

The small blob represents the volume of space being scrutinized by Project Phoenix. The new Allen Telescope Array will be able to examine the far larger chunk of space indicated by the orange blob.

Fair enough; we seem to have this part of the Milky Way to ourselves. But just how far is it to the nearest active alien civilization?

Obviously, we still don't know. It depends on how many societies populate the Galaxy. This is the number *N* in the Drake Equation. The pessimists would say that *N* is one; in other words, we're the only game in town. In that case, the nearest aliens are . . . well, there aren't any nearest aliens! Astronomer Carl Sagan, on the other hand, was a lot more optimistic. He once suggested that *N* might be a million. In that case, it's easy to work out that the average distance to the nearest galactic civilization is roughly 100 light-years. That's comparable to the amount of interstellar space we've actually searched.

Drake himself is inclined to a more modest estimate for N, namely that there are approximately 10 thousand technological civilizations sprinkled throughout the Milky Way. If he's right, then our nearest neighbors will be 500 to 1,000 light-years distant. That's a fair piece, and rather farther than our most sensitive searches have reached, although it's only one percent of the distance across the Galaxy.

The plain and simple fact is that we won't know the mileage to ET's home until we find a signal. But unless you're of the opinion that the Galaxy fizzes with more than a million inhabited worlds, the distance will be 100 light-years or more. Conversation – should we ever attempt it – will be mighty slow; centuries for a query and a reply, minimum.

But that's OK. In the sixteenth century, communication between Europe and the newly discovered Americas was also poky. However, the slow speed of information exchange didn't make the discovery less important. Just knowing there were societies on the other side of a vast sea was reason enough to stop and wonder.

But if interstellar travel happens at all, then one star system could seed others. The number of transmitting societies might be large, even if the chance of producing intelligence is small. It's just that one society could produce many others.

Another point is that in discussing the equation, we've limited ourselves to biological intelligence. As noted earlier, it may be that our destiny is to invent machine intelligence, and sentient circuitry will inherit the Earth. If so, these ultrasmart machines might eventually abandon our planet for greener pastures. This scenario may have already played out elsewhere. The picture of a galaxy peppered with thinking machines, with only occasional sites of limited and fragile biological intelligence, is quite different from the one painted by Drake's equation.

Finally, there's the possibility that the equation overlooks the fact that Earth might be very special. Perhaps the emergence of complex life depends on circumstances that are exceedingly unlikely. Earth, after all, is not only in the right part of the Solar System to have tolerable temperatures and liquid water, but we also have a large moon (rare for a rocky planet) that helps to stabilize our planet's daily spin. In addition, the presence of Jupiter has helped to clean out our Solar System of most of the large rocks that otherwise would often slam into our planet and wreak

Our nearest, large neighbor galaxy, the Andromeda Nebula.

Smart machines might occupy the depths of space.

This image of our Moon was taken by the Galileo spacecraft as it flew past.

The dark splotches are the result of the bombardment of the giant planet Jupiter by comet Shoemaker–Levy 9 in 1994. Better that Jupiter take the hit and not Earth!

havoc and destruction on scales Godzilla could only envy. All these circumstances and more have been cited to suggest that Earth is enormously more special than the most ardent of environmentalists would claim, and that f_L and/or f_i are essentially zero, making the equation meaningless.

Drake himself finds such discussions interesting and of occasional value. But in the end, the only way to decide if there is other cosmic intelligence – be it protoplasmic or engineered – is to build the telescopes and search for it. We're now doing that, and new instruments coming on-line in the next decade will dwarf the SETI experiments of today. So if you are an optimistic person, you might reasonably expect that signal detection could occur within your lifetime. What, precisely, would that mean?

CHAPTER 7

The future

So far, no SETI experiment has ever found and confirmed a signal from another world. Consequently, it's hardly surprising that one of the questions most frequently put to the researchers is, "don't you get discouraged?" After all, it's been more than four decades since Project Ozma. Even the most patient explorers might be expected to balk after four decades of unrewarded searching. Columbus – or at least his crew – turned cranky after only a few *weeks* of tossing on the Atlantic.

But unlike Columbus' feckless swabbies, SETI scientists are not on the verge of mutiny (although to be fair, rather few of them are subsisting on hardtack and bilge water). "Sail on," they say. This upbeat attitude is more than uncommon patience and blind hope. Recent developments in astronomy, as well as in SETI itself, have spurred their quest, urging them to not only continue the search, but to expand it.

A promising situation

Why the upbeat attitude? Scientists will readily confess that, when push comes to academic shove, there's still no fully convincing proof of life beyond Earth – intelligent or otherwise. But one thing that *has* been proved is that planets are plentiful. That became clear beginning in 1995, with the discovery of a Jupiter-size world around the nearby star, 51 Pegasi. Since then, researchers have found that at least one in ten stars has a single planetary companion or more, which means there are tens of billions of planets (at a minimum) hiding in the dark depths of the Milky Way.

That's good news on the astronomy scorecard. But there's also suggestive action on the biology front. We've already mentioned the 1996 claim by NASA scientists and others that fossilized microbes are hunkered inside a meteorite known to have come from Mars. This meteorite has developed so much notoriety that its name – ALH 84001 – is probably more frequently recognized by astrobiologists than the names of their first-born.

But notoriety aside, the claim that Mars once had tiny inhabitants has been hotly disputed. Skeptics maintain that the microscopic "fossils" visible in the meteorite are completely natural mineral structures that have nothing to do with life. Rebutting these arguments, the discovery team points to miniscule crystals of magnetic iron (magnetite) that they've found within ALH 84001. These crystals, they say, could only have come from microbes. They bolster their claim by pointing to certain earthly bacteria which produce magnetite that functions as a compass needle to help them navigate in an ocean environment. But of course, magnetite can be formed by geology as well as biology. Are those microscopic compass needles merely geologic accidents? Biologically produced magnetite has properties that are subtly different from those of its geologic cousin, so the debate over fossils in ALH 84001 now turns on the precise properties of peewee grains of iron.

The arguments continue, but the bright side of all this – at least from the SETI researchers' perspective – is that there's actual evidence for possible life from space. In 1976, when the robotic Viking Landers scraped up some Martian dirt in their search for bacterial beings, they failed to find even simple organic molecules. This convinced the biology team back on Earth that, despite centuries of optimistic speculation about Martians, the Red Planet was as dead as mutton. A mere two decades later, reputable scientists are claiming that this rusty world once had life (and are quick to add that it may *still* harbor microbes under its sterile surface). So the prospect of life on Mars, at least, is once again respectable.

We've also mentioned the stunning possibility that several of Jupiter's moons might have jumbo-size oceans beneath their frozen skins. In particular, there is persuasive evidence that Europa – a satellite very nearly the same size as our own Moon – is hiding a salty, world-girdling ocean containing twice as much water as all the seven seas of Earth. We can thank the Galileo space probe for most of this evidence. Galileo began its reconnaissance of the jovian system at the end of 1995. Its high-resolution photos of Europa show kilometer-size blocks of ice that seem to be locked into a solidified sea. The fact that the blocks are often widely separated suggests that liquid water occasionally seeps up from underneath the ice, causing it to break into bergs. More convincing evidence for a subsurface

Viking took this picture in 1976 which showed Mars and no Martians . . . unless they are cleverly disguised as rocks.

ocean is the space probe's measurements of Europa's slight magnetism. The magnetism varies as the moon orbits Jupiter, and the changes suggest that hidden salt water is acting like a coil of wire, mimicking a moon-size dynamo as Europa moves through Jupiter's strong magnetic field.

So the reason that SETI scientists continue to smile is that during the past decade, the research winds have all blown in a favorable direction: suggesting a Universe filled with worlds, and one in which biology can spring up even on less-than-perfect planets or unremarkable moons. NASA and space agencies in Europe and Japan are moving as quickly as their budgets will allow to reexamine Mars, and to probe the enticing moons of Jupiter and Saturn. If even *one* of these nearby locales is found to actually have life – even simple life – then we will confront the virtually inescapable

Galileo approaching Jupiter's ice-covered moon, Europa.

conclusion that biology is as common as phone poles. Obviously, that will be a great impetus to SETI researchers.

Better telescopes

Since new discoveries are consistent with the possibility that life (including clever life) might be bubbling away on countless worlds, the SETI community is busy trying to improve its experiments.

Currently, the most comprehensive search for extraterrestrial radio signals is Project Phoenix. As we noted in Chapter 5, this experiment is examining a thousand nearby star systems, and doing so over a greater part of the radio dial, and with greater sensitivity, than any previous SETI search. That sounds encouraging. But note that a thousand stars – comparable to the total number recognized by the Greeks and Romans – is, in fact, merely a drop in the galactic bucket. The mighty Milky Way has hundreds of *billions* of stars. Put it another way, if you were searching Africa for big game, this would be like checking out a patch of savannah the size of a city block. Hardly a thorough search.

Even worse is the fact that the experiment is slow. Intensive SETI searches burn up valuable telescope time, and Project Phoenix is fortunate to get several weeks on the Arecibo Radio Telescope each year. But the telescope can point to only one star system at a time, and even then it takes about eight hours of observing to examine the nearly two billion channels that Project Phoenix monitors. Add it all together, and this experiment – which is enormously better than its predecessors – plows through its target list at a speed of only 50–100 stars per year. At that rate of stellar scrutiny, a search of even 1 percent of the Milky Way's stars would take *millions* of years. Clearly this would demand dedication of a high order.

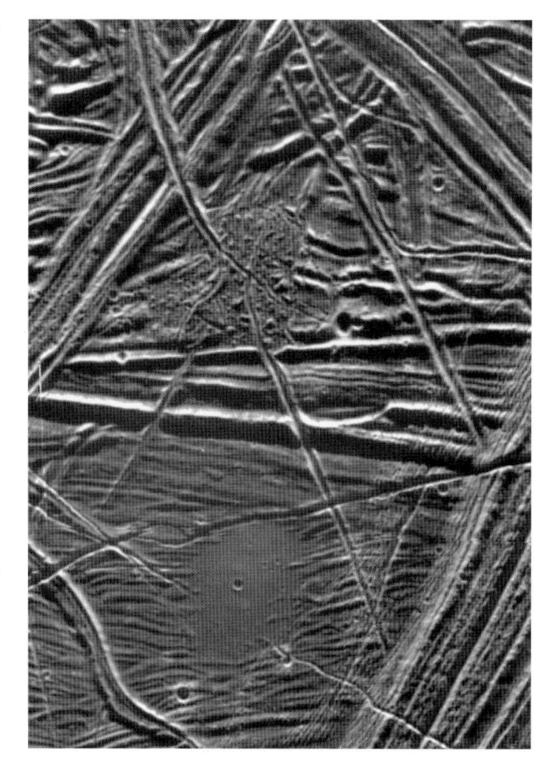

Close-ups of Europa, showing its cracked and crazed, hard ice surface. There is suggestive evidence that 10–15 km beneath this ice crust is a vast and deep liquid ocean.

So the first thing that SETI scientists are keen to do is to speed up the search, while somehow maintaining sensitivity and spectral (bandwidth) coverage. To this end, the SETI Institute has joined with the University of California at Berkeley to construct a new type of radio telescope. Instead of a huge, single dish antenna, such as the one at Arecibo, this new telescope will be comprised of hundreds of smaller antennas. In the parlance of astronomers, it will be an array. Radio astronomers already field such multi-element telescopes (the Very Large Array, or VLA, in New Mexico, is sufficiently photogenic to be a frequent prop in movies, including *Contact*), but like their large, single-dish brethren, these antenna teams are constructed of custom-built components that are large (typically 25 meters in diameter) and costly.

The SETI Institute's new instrument – christened the Allen Telescope Array (after Paul G. Allen, a software pioneer who has donated much of

The Very Large Array is not the largest radio telescope, but it is probably the best known thanks to its starring role in several movies.

the money to build it) – will be constructed from relatively inexpensive, smaller antennas. These are modified backyard satellite dishes, 6 meters in diameter, fitted with motors that will allow them to hold a fix on a particular patch of sky as the Earth rotates. The new array will be situated at Hat Creek, a small, rural town in the mountains 500 km northeast of San Francisco. This lonely location suffers less of the human-made radio interference that flusters radio telescopes in more urban areas. When the telescope is completed, 350 of these souped-up satellite dishes will be spread over roughly a square kilometer of landscape – a veritable antenna orchard.

What's the advantage of making a telescope from small antennas? To begin with, it's considerably less costly, since satellite dishes are a commodity

A computer-generated model of the Allen Telescope Array from the air. This new instrument will consist of 350 antennas, each 6 m in diameter.

product, fabricated by the millions. Sure, they have to be modified to make them better suited for studying the sky, but they remain inexpensive. The second advantage of an array is its ability to observe fine detail. Physics assures us that bigger antennas can "see" smaller details. An array spread over a kilometer of turf can achieve the same magnification as a single antenna a kilometer across. For the University of California radio astronomers who will be sharing use of the Allen Telescope Array, this ability to map fine structure in a gas cloud or a galaxy is important.

But there's another advantage of an array that is of direct benefit to its SETI work: it can look at more than one star simultaneously. The Arecibo

The mother of all radio telescopes

When it comes to telescopes, bigger is almost always better.

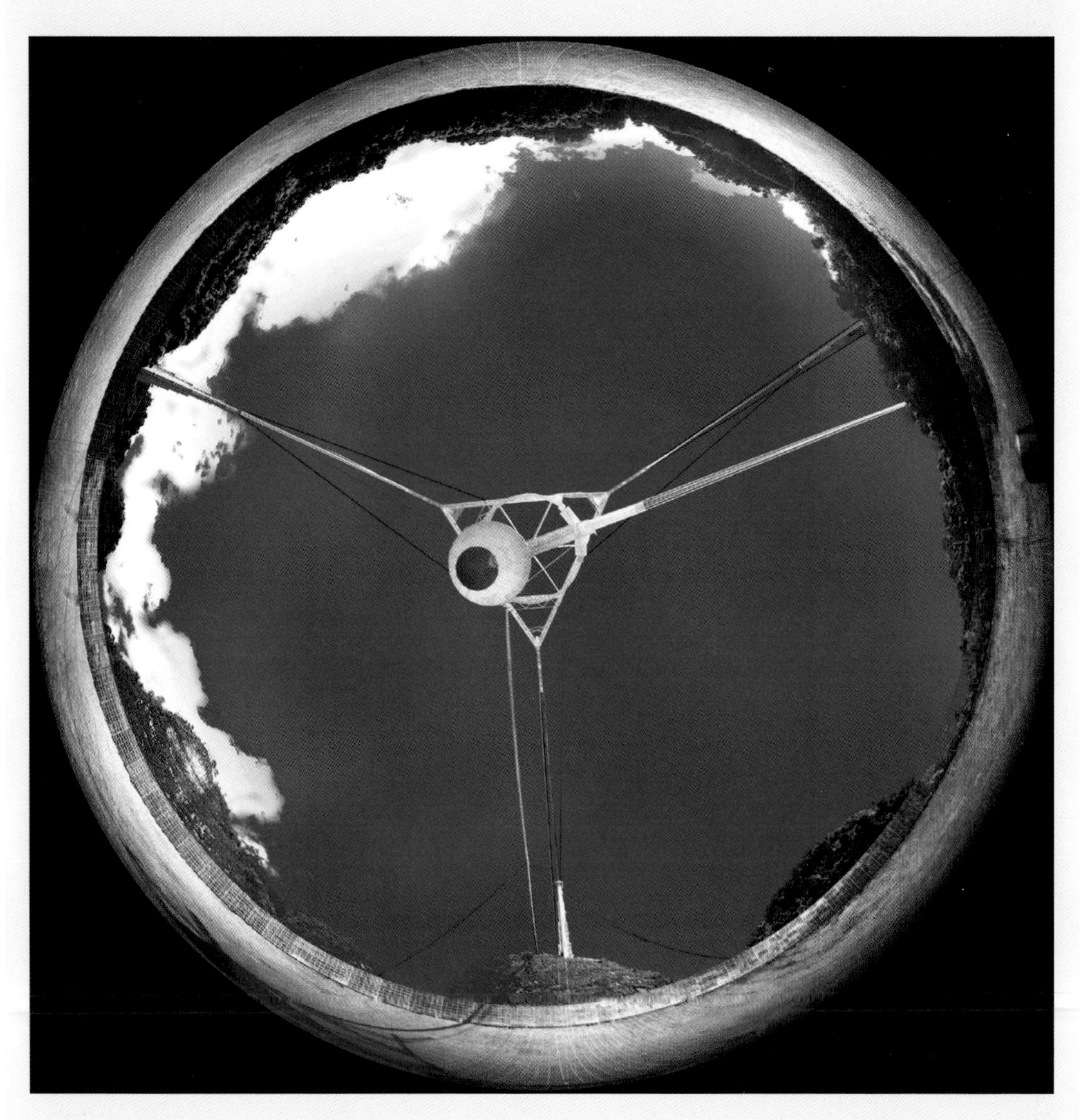

An unusual view of the Arecibo Radio Telescope taken from the centre of the 305-m dish looking up to the receivers.

The Australian SKA concept uses an unusual radio "lens" called a Lunaberg lens.

That's because, from the astronomers' point of view, the most interesting objects to study are often the farthest away. For example, most curious folks are keen to know more about the expansion of the Universe. Learning this would tell us a great deal about our future fate, and when we can expect a final cosmic fade to black.

But solving this mind-boggling puzzle requires peering back in time, in order to measure the expansion when the universe was young. For astronomers, looking into the past can be done by looking far away: we see objects that are five billion light-years distant as they were when the Cosmos was five billion years younger. So like Indian scouts on a high butte, astronomers are always straining to see farther.

The Dutch proposal for the SKA involves a large number of flat antennas with built-in receivers.

The Canadian design uses an antenna tethered by balloon above a huge reflector.

But sadly, farther usually means fainter. Distant galaxies are millions of times dimmer than those nearby, and that's why large telescopes – both optical and radio – are always in great demand.

The radio telescope sporting the greatest collecting area is the 305-meter Arecibo dish, in Puerto Rico. This brobdingnagian beast is used to study the cosmic static from distant galaxies, but it's also the instrument wielded by scientists for two major SETI experiments, Project Phoenix and SERENDIP IV.

It's been four decades since Arecibo was built – a long time to hold the heavyweight title. But plans are starting to take shape for a far larger radio instrument. Called the SKA, or Square Kilometer Array, it will have ten times the collecting area of Arecibo, and could probe objects three times farther off. Since the SKA will be able to run several experiments simultaneously, it could be used for "piggyback" SETI programs. The advantages are considerable: it could find transmitters at three times the distance possible with Arecibo, or detect nearby transmitters that are far below today's sensitivity threshold.

The idea for the SKA came out of a 1994 study by radio astronomers interested in new instrumentation. Today, it's being worked on by an international consortium of scientists, and this worldwide cooperation will undoubtedly extend to the SKA's eventual funding and construction. At the moment, it's still not clear exactly what form this massive instrument will take. One suggestion is to simply build ten Arecibo-style dishes, and operate them as a team. Those familiar with the maintenance requirements of the Puerto Rican telescope have been known to blanch at this idea. Another possibility is to make one very large, very shallow reflector (which is relatively simple to build), and then support the receiver at the focus with tethered balloons. Still another approach is to build up the SKA as a family-size version of the Allen Telescope Array, using customized backyard satellite dishes. The number of dishes required would be tens of thousands, but this scheme could easily prove to be the most attractive. One reason why the construction of the Allen Telescope Array is being so avidly watched by the astronomical community is because it's an important test-bed for technology useful for the SKA.

A decision about which technological approach to take for the SKA will be made in 2005, but construction won't begin until after 2007. Once up and running, it will be the unchallenged mother of all radio telescopes, with more collecting area than a dozen of its rivals.

telescope is like a digital camera with only one (count 'em, *one*) pixel. You aim it at a single target at a time. But the data from an antenna array can be combined in such a way as to form many pixels on the sky at once – a few, a few dozen, or a few hundred (how many depends primarily on the available computer power). So in practice, when the Allen Telescope Array is aimed by the Berkeley radio astronomers at some interesting target to make a "radio picture," the SETI researchers will be using a handful of pixels to examine nearby stars that happen to be in the same part of the sky. Dual use means that SETI can collect data 24 hours a day, 7 days a week. This non-stop observing, coupled with the ability to scrutinize several star systems simultaneously, means they can check out cosmic neighborhoods far faster than now.

The bottom line is that, from the day it's turned on, the Allen Telescope Array will investigate nearby stellar habitats at an annual rate more than 100 times faster than has been available so far.

It's not only radio SETI that's planning speedier new search hardware. Like their radio counterparts, the experiments designed to hunt for flashing laser lights from other worlds are also examining only a single star system at a time. But new, high-speed, multi-pixel photo detectors are about to up the pace. A 1.5-meter telescope being built by Harvard University will slowly

Gordon Moore, the man behind Moore's Law.

scan the entire northern sky, hunting for pulses of light that might be an interstellar message. If a flash is found, it can be quickly checked with a similar instrument located at Princeton University, about 500 km southwest of Harvard. This geographic separation is important for the experiment. Depending on where the telescopes are pointing, light from the target star system will generally have to travel a few hundred kilometers farther to one telescope than to the other. You might think this is insignificant, given the fact that the light has already traversed many trillions of kilometers. But the small extra distance means that the flash would typically reach one telescope a thousandth of a second before it reached the other (light takes 0.0003 seconds to go 100 km). The slight delay is easily measured, and is an excellent means of being sure that the light pulse is coming from the star being observed, and is not merely a flash caused by instrumental malfunction or nasty cosmic rays.

SETI is getting better, and much of the improvement is due to the relentless march of technology. It's not just *any* technology that's pushing SETI; it's the rapid development of digital electronics. Gordon Moore, a well-known Silicon Valley pioneer, noted decades ago that the number of transistors that can be fabricated on a computer chip doubles every two years. The truth is that technological progress has accelerated, and today computer chips improve by a factor of two every 18 months. (This is the main reason why your home computer so quickly becomes obsolete.) Since much of the speed at which a SETI experiment can probe the sky depends on the digital equipment behind the telescope, a doubling of computing power usually results in a doubling of search power.

An instrument such as the new Allen Telescope Array is particularly suited to take advantage of this relentless technological uptick. If it can peruse 1,000 star systems per year when completed, then a decade later it should be capable of inspecting 100,000. Assuming that the pace of technology is sustained, the total number of star systems canvassed by this telescope will reach millions by the year 2025.

That's an interesting thought, and here's why. In the last chapter, we remarked that SETI pioneer Frank Drake reckons that the number of transmitting civilizations in our Galaxy is about 10,000. If he's right, then roughly one Sun-like star in a million has intelligent inhabitants. One in a million. If our SETI strategy is to scout out Sun-like stars for signals, we'll need to survey a million or so to have a good chance of finding our first alien society. But as we see, that could happen by 2025.

Despite 40 years of searching, SETI experiments have yet to find a peep or a flash from the depths of space that's clearly due to alien intelligence. This has led some folks to somberly intone that it could be centuries or longer before the hunt pays off (if it *ever* does). But the exponential growth in digital electronics suggests otherwise. We could hear ET within your lifetime.

And what if we did?

Champagne in the Arecibo observing-room fridge – just in case!

The consequences of contact

As we described in Chapter 5, detection and confirmation of a signal will take several days. This is rather different from the situation in Hollywood's fictional portrayals. In the movies, scientists who have spent years listening to the endless, mindless rumble of the Cosmos suddenly light up when a weird signal breaks through the static. Within minutes, there's complete mayhem. It's a "Eureka!" moment. Soon, everyone in the world knows about the discovery, and more often than not, the government steps in and takes over.

It sounds and looks believable. But is that the way it would really happen?

Well, no. To begin with, just picking up a weird-sounding signal is no guarantee that it's alien. Plenty of terrestrial radar signals sound "weird."

To actually be sure that a signal is coming from a distant star system, and not just the local airport or a telecommunication satellite wheeling overhead, will require careful checking, a process that could take the better part of a week.

No "Eureka" moment, in other words. But throughout the verification efforts, you can be sure that the scientists involved will be telling friends and relatives about the "interesting signal" they're working on. And those folks will, inevitably, tell the world at large. Experience with false alarms shows that the media rapidly learn of such events, and immediately inquire as to what's going on.

So one thing that's quite certain: the news of a detection will be splattered across newspapers and TV screens even while the discoverers are still busy convincing themselves that the signal is real. According to an informal poll made of journalists a few years ago, finding an extraterrestrial signal would rank among the biggest news stories of all time.

What would be the reaction of the public? It's unlikely that there would be widespread fear or panic, although that occasionally happens in fiction. Keep in mind that a large percentage of the population already believes that aliens are here on Earth. Despite this supposed alien presence, these people are not rioting in the streets (at least, they're not rioting in the streets about extraterrestrials). They confine their alien anxieties to badgering radio talk show hosts about getting the government to "come clean" about UFOs and crashed saucers. The dead-simple fact is that a signal from hundreds or thousands of light-years distance poses no threat to us whatever. The alien broadcasters won't even know that we've picked up their transmission.

But, you might wonder, what if we reply? What if we acknowledge their transmission with something like, "We hear you. Here's some information about our world," and accompany the broadcast with pictures, music, and anything else that somebody thinks is worthwhile about humans. That sounds reasonable, but maybe not. Perhaps answering the phone would be a *bad* idea, since it would tip off the extraterrestrials that someone's on this planet, possibly encouraging them to visit. But here's a problem: we've *already* replied. Our TV and radar signals are dozens of light-years into space, irrevocably signaling our existence to other star systems. Some concerned citizens have even suggested that these signals – motivated primarily by humankind's wish to sell toothpaste and bath soap – will soon provoke the Apocalypse. They postulate that advanced societies are carefully and silently monitoring likely star systems, awaiting a signal that would indicate that yet another world has finally developed the ability to build destructive and competing hardware. Once they detect a signal, these folks say, the listening societies will launch their missiles, and destroy this new, and potentially threatening civilization. So in their sunny view, it's just a matter of time before robotic weapons sent our way by defensive aliens, replying to threatening deodorant advertisements, wipe us out. If this is true, then you should consider being kind to the neighbors, as we're all doomed. But it won't have been SETI or a detected signal that provoked this bizarre demise: it was simply the emergence of radio technology.

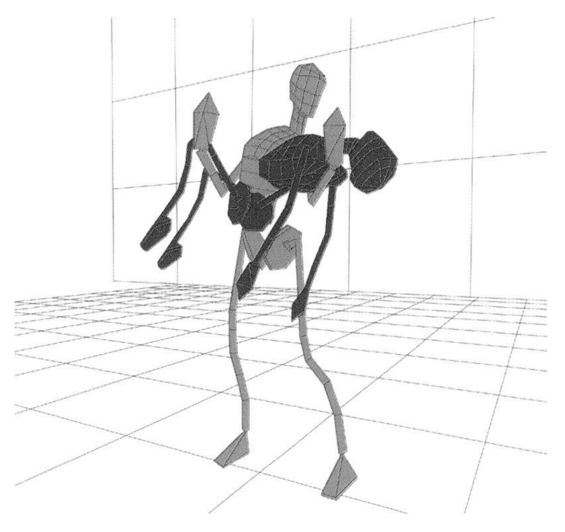

This image, developed as part of a study of interstellar communication, represents altruism, but could it be anything else? If we or the aliens use pictures to communicate, will that help to bridge the culture and language gap?

Paranoia aside, the most immediate question that's likely to be asked once a signal is found is not "should we reply?" but rather "what are they saying?" The answer could be a long time coming. If the incoming signal consists of flashing laser pulses, then it can be quickly written to a computer memory, and later distributed to anyone with a flair for cryptography. If the signal comes in via a radio telescope, it's likely that the "message" will be lost. That's because radio SETI experiments average incoming data for several seconds or minutes. They do this to build up sensitivity, in the same way that a time-exposure photograph builds up sensitivity to faintly lit scenes. But in doing so, the radio experiments throw away any part of the signal that varies rapidly – and that will almost surely include the message. If we detect aliens with our radio telescopes, we'll know they're on the air, but we won't get the programming.

The obvious solution to that problem is to build a much larger, more sensitive radio telescope that *could* find the message. And once again, when the message is found, it would surely be distributed to as many people as possible for deciphering. Considerable work has been expended in thinking

What if we don't get a signal?

Thanks to the relentless march of technology, today's radio SETI experiments are reckoned to be 100 trillion times more effective than the first such search in 1960.

And yet, we still haven't heard the aliens. What happens if many more decades go by, and we *still* haven't detected extraterrestrial signals? Would researchers be willing to throw in the proverbial bath linen, and admit "they're just not out there"?

That seems to depend on whom you ask. Many SETI scientists will note that anyone setting out to comb the Galaxy for inhabitants needs to be exceedingly patient. It's a quest, they say, that could take many generations, and we'd best be prepared for the long haul. It's much too early to get upset about SETI's failure to find a signal.

On the other hand, not everyone agrees that success is merely a matter of continuing the search. In 1950, the Italian-American physicist Enrico Fermi asked during lunch "where is everybody?" This wasn't a reference to missing dining companions, but to a profound thought. Fermi reckoned that if intelligence springs up often in the Galaxy, then some beings would have long ago mastered space travel. By now they should have inhabited every half-decent star system. But *we don't see* the aliens in our neighborhood (unless you're inclined to believe that UFOs are extraterrestrial flying machines). So there's a problem here: we suspect that alien intelligence has probably developed on many worlds, and yet they don't seem to have colonized our part of the Galaxy.

This small puzzle is called the Fermi Paradox. Clearly, if SETI continues to make millions of fruitless scans of the sky, there will be some folks who will claim that Fermi's remark should be taken seriously. We don't see the aliens here, because they're not *there*. SETI isn't finding anything because there's nothing to find.

Of course, that's an argument that few SETI scientists would accept. Just because we don't find a signal doesn't mean that we're alone in the Cosmos. "Absence of evidence is not evidence of absence," as the SETI researchers are quick to note on talk shows and at dinner parties. Perhaps there are methods of communication that are so much better than radio or light that all advanced societies are using those, and we are just naively pursuing the wrong search strategy. This is a reason why SETI researchers are always discussing new ways to look. If you think about it, it's very difficult to prove that we *are* alone in the Universe.

Still, everyone who's looked carefully at the amount of stars that we've scrutinized admits that it's much too early to draw any conclusions about whether there are transmitting aliens amongst the star fields of our Galaxy. The number of stellar habitats that have been carefully checked out is only a few hundred. As a reminder, there are some hundreds of billions of stars in the Milky Way. It's far too soon to think of quitting. Columbus didn't turn back after only a few hours sail.

about how alien messages might be encoded: after all, we won't know their language, and perhaps they don't even *have* language. Perhaps they'll send us pictures. They might try to encode their message using music, or the universal language of mathematics. All of this is possible, but we won't know which scheme – if any – is actually in use until we get a signal.

Keep in mind that the transmitting society will be far away. Dozens of light-years at a minimum, but more likely they'll be hundreds or even a thousand light-years from Earth. Two-way communication will be tedious, a fact that will be well known to the alien broadcasters. Therefore, it's reasonable to assume that their broadcast will be *de facto* one-way. They'll be giving us a message, not initiating a conversation. This has led some SETI scientists to postulate that – since the senders will very likely be far in advance of us (they have to be at least as sophisticated as we are to send the signal at all!) – they might wish to educate us by giving us their encyclopedia. This is the practical up side to SETI. They might not only tell us we've got cosmic company, but might also provide us with knowledge that otherwise would take us centuries to discover. But even if they've sent nothing of consequence, or nothing that we can ever decipher, the very fact that an advanced society exists would not only be profoundly important in a philosophical sense. It would also be good news, since it would prove that not all technologically adept societies inevitably self-destruct.

Of course, and as we've mentioned, it may be that an altruistic message from sentient aliens is not what we'll pick up at all. We should be prepared for the possibility that any signal we receive was transmitted, not by a superior and supple race of biological beings eager to spread enlightenment throughout the indifferent darkness of the Galaxy, but by the intellectually superior products of their technology: thinking machines. It could be that all we will hear are impenetrable data streams shuffling between complex neurocomputers – the true intellects of the Cosmos.

But whether the signal comes from biological beings or solid-state sentience, a SETI detection will tell us something that will forever change the way we view ourselves: namely that we are only a single tile in a vast, cosmic mosaic. The barrier of isolation that has separated our planet from the Universe for billions of years, that has kept our eyes down and our thoughts parochial, will shatter. The world will change overnight, and we will wake to find that all our histories, and all our stories, are merely a small entry in an enormous book.

Further reading

Andreas, A. *To Seek Out New Life: The Biology of Star Trek*, Crown, 1998
Ashpole, E. *The Search for Extraterrestrial Intelligence*, Blandford, 1990
Bennett, J., Shostak, S. and Jakowski, B. *Life in the Universe*, Addison-Wesley, 2002
Bracewell, R. N. *The Galactic Club: Intelligent Life in Outer Space*, San Francisco Book Company, 1976
Cameron, A. G. W. *Interstellar Communication*, W. A. Benjamin, 1963
Cohen, J., and Stewart, I. *What Does a Martian Look Like? The Science of Extraterrestrial Life*, John Wiley and Sons, 2002
Darling, D. *The Extraterrestrial Encyclopedia: An Alphabetical Reference to All Life in the Universe*, Three Rivers Press, 2000
Life Everywhere: The Maverick Science of Astrobiology, Basic Books, 2001
Davies, P. *Are We Alone? Philosophical Implications of the Discovery of Extraterrestrial Life*, Basic Books, 1996
Dick, S. J. *Life on Other Worlds*, Cambridge University Press, 1998
Dickinson, T. and Schaller, A. *Extraterrestrials: A Field Guide for Earthlings*, Camden House Publishing, 1994
Drake, F. and Sobel, D. *Is Anyone Out There?*, Delacorte Press, 1992
Ekers, R. *et al.*, eds., *SETI 2020: A Roadmap for the Search for Extraterrestrial Intelligence*, SETI Press, 2002 [technical description of the future of SETI]
Fichtman, F. *SETI*, Penguin Books, 1990
Fisher, D. E. and Fisher, M. J. 1998, *Strangers in the Night: A Brief History of Life on Other Worlds*, Counterpoint Press, 1998
Goldsmith, D. and Owen, T. *The Search for Life in the Universe*, University Science Books, 2001
Harrison, A. *After Contact: The Response to Extraterrestrial Life*, Plenum Press, 1997
Jackson, E. *Looking for Life in the Universe*, Houghton Mifflin, 2002 [for ages 9–12]
Mallove, E. and Matloff, G. *The Starflight Handbook*, John Wiley and Sons, 1989
McConnell, B. *Beyond Contact: A Guide to SETI and Communicating with Alien Civilizations*, O'Reilly and Associates, 2001
Levay, S. and Koerner, D. *Here Be Dragons: The Scientific Quest for Extraterrestrial Life*, Oxford University Press, 2002

O'Neill, Gerard K. *The High Frontier: Human Colonies in Space*, William Morrow, 1977
Parker, B. R. *Alien Life: The Search for Extraterrestrials and Beyond*, Plenum Press, 1998
Regis, E., Jr., ed. *Extraterrestrials: Science and Alien Intelligence*, Cambridge University Press, 1985
Schmidt, S. *Aliens and Alien Societies*, Writer's Digest Books, 1995
Shostak, S. *Sharing the Universe: Perspectives on Extraterrestrial Life*, Berkeley Hills Books, 1998
Skhlovskii, I. S. and Sagan, C. *Intelligent Life in the Universe*, Holden-Day, 1966
Sullivan, W. *We Are Not Alone*, Penguin Books, 1993
Ward, P. and Brownlee, D. *Rare Earth: Why Complex Life is Uncommon in the Universe*, Copernicus Books, 2000
Webb, S. *If the Universe Is Teeming with Aliens... Where Is Everybody? Fifty Solutions to Fermi's Paradox and the Problem of Extraterrestrial Life*, Copernicus Books, 2002

Image credits

Seth Shostak: 2, 4, 32t, 39t, 45, 46, 48, 49, 62, 66, 75, 78, 84, 97, 103, 104, 112–113, 114b, 116, 121, 128, 136, 142, 151; Seth Shostak/NASA: 93; National Space Centre: 3, 5, 18–19, 20–21, 23, 25, 31, 38, 40–41, 51, 52, 53t, 53b, 54, 55, 56t, 56b, 63, 68, 69, 71, 73(all), 76, 80, 81, 90, 117, 146; R. Williams and the HDF Team/NASA/STScI: 9; SOHO/EIT Instrument: 11; D. Padgett (IPAC/Caltech), W. Bradner (IPAC), K. Stapelfeldt (JPL)/NASA: 12; NASA/Hubble Heritage Team: 13; N. A. Sharp, NOAO/NSO/Kitt Peak/FTS/AURA/NSF: 14; HST/Jon Morse (University of Colorado)/NASA: 15; ESA 2001 Medialab: 16, 22; NASA/JSC: 24; NASA/JPL, Craig Altebery: 27; NASA: 28, 32b, 88, 89, 92, 106, 130, 136b, 143; ESA/Beagle 2, All rights reserved: 29; ESA Double Cluster Image, A. Steere: 33; J. Caldwell (York University, Ontario), Alex Storrs (STScI)/NASA: 39b; NASA/JPL/University of Arizona: 47; NASA/Pat Rawlins: 91; NASA/JPL: 94, 141; NASA/JPL/Malin Space Science Systems: 95; SETI League/P. Shuch, Photo used by permission: 100t, 100b; National Radio Astronomy Observatory (NRAO): 102t; SETI Institute: 102b, 105b, 109; Seth Shostak/SETI Institute: 105t, 107, 110, 114t, 112; Courtesy Big Ear Observatory: 108t; Courtesy SOHO (Solar and Heliospheric Observatory): 108b; Oak Ridge Observatory/Harvard: 111; NASA/STScI: 123, 127; National Space Centre/Blue Streak, Courtesy National Museums and Galleries on Merseyside (Liverpool Museum): 129; Andy Levin/Parade, Courtesy of the Estate of Carl Sagan: 132; Ly Ly/SETI Institute: 133; Bill Schoening, Vanessa Harvey/REU Program/NOAO/AURA/NSF: 135; R. Evans, J. Trauger, H. Hammel and the HST Comet Science Team: 136b; Image courtesy NRAO/AUI: 144; Isaac Gray/SETI Institute: 145; C. Fluke, Center for Astrophysics and Supercomputing, Swinburne University of Technology: 147; ASTRON, the Netherlands: 148t; National Research Council Canada Dominion Radio Astrophysical Observatory: 148b; Intel Corporation: 150; Doug Vakoch/SETI Institute: 153.

Index